Arthur Nicols

The Puzzle of Life; and How it Has Been Put Together

A Short History of the Formation of the Earth, with its Vegetable and...

Arthur Nicols

The Puzzle of Life; and How it Has Been Put Together
A Short History of the Formation of the Earth, with its Vegetable and...

ISBN/EAN: 9783337095321

Printed in Europe, USA, Canada, Australia, Japan

Cover: Foto ©berggeist007 / pixelio.de

More available books at **www.hansebooks.com**

THE PUZZLE OF LIFE;

AND

HOW IT HAS BEEN PUT TOGETHER.

A SHORT HISTORY OF THE FORMATION OF THE EARTH,
WITH ITS VEGETABLE AND ANIMAL LIFE,
FROM THE EARLIEST TIMES,

INCLUDING AN ACCOUNT OF

PRE-HISTORIC MAN, his WEAPONS, TOOLS, and WORKS.

BY

ARTHUR NICOLS, F.R.G.S.

WITH ILLUSTRATIONS by FREDERICK WADDY.

SECOND EDITION.

LONDON:
LONGMANS, GREEN, AND CO.
1877.

All rights reserved.

TO

MY YOUNG FRIENDS

BEATRIX, GUY, SYLVIA, MAY, AND GERALD,

THE CHILDREN OF

GEORGE DU MAURIER.

PREFACE

TO

THE SECOND EDITION.

THE favourable reception accorded to the first edition has induced me to give the present a more definite educational character. Foot-notes are appended, referring to the position in the British Museum of all the principal antiquities, fossils, and implements mentioned in the text; so that the specimens can easily be found by any young student who wishes, with the book in his hand, to make himself familiar with these records of past time. This will probably facilitate the search for and recognition of specimens by the reader.

PREFACE TO THE SECOND EDITION.

The additions to the text consist chiefly of a more extended account of the deposition of chalk and other deep-sea formations, founded on the results of the "Challenger" and "Tuscarora" expeditions, and a sketch of the earth-works of the Ohio mound-builders and the stone monuments of Easter Island. Examples of pre-historic art and lake-dwellings have been added to the illustrations.

A. N.

HAMPSTEAD: *March* 1877.

PREFACE

TO

THE FIRST EDITION.

HAVING found that children could be interested in the history of life upon the Earth, and that it appealed forcibly to their understanding, I considered that a little book upon the subject might give them the taste for more extended study in after years. The difficulty of treating the, to them, novel conclusions of geology, often founded on abstract reasoning, in language simple in form yet stating clearly the great principles upon which this reasoning rests, will probably be apparent on every page. Breadth, rather than minuteness, has been aimed at, in the belief that a general view, not overcrowded with details, is likely to be the most impressive. Thus, in the geological part the leading features of the succession of strata have been preserved, but no details of systematic classification entered into. Similarly, Primeval Man is considered mainly with reference to gradual progress from a rude to a more civilized condition. To have been more explicit, where there is still much difference of opinion, would have obscured the main facts of the evidence for man's great antiquity.

The illustrations are typical examples of the three arbitrary but convenient divisions of the history of life—the vegetable, the animal, and the human—such as will be most readily met with in museums. Slight as this sketch is, the liking for it shown by some intelligent children, who saw it in manuscript, encouraged me to believe that there are many others to whom it might prove interesting.

Some acquaintance with the leading facts in science is daily becoming more necessary to those who aspire to liberal culture, and instruction in them is a recognised feature in the curriculum of some public and leading private schools. Thus, it is hoped that the present volume may to some extent serve as a text-book without the severity of such a form. The best English and foreign authorities have been consulted, and other trustworthy sources —as papers read before scientific societies— drawn upon, bringing the information down to the latest time. Though these pages are designed for young persons, other readers, perhaps, who are not familiar with the subject, may find some interest in them if they are not deterred by the necessarily simple style.

My thanks are due to Mr. H. B. WOOD-WARD, of the Geological Survey of England and Wales, for some valuable suggestions made during the progress of the work.

A. N.

HAMPSTEAD : *November* 1876.

CONTENTS.

	PAGE
THE FRAMEWORK OF THE PUZZLE	1
THE GEOLOGICAL PART	17
THE VEGETABLE PART	56
THE ANIMAL PART	77
THE HUMAN PART	120
CONCLUSION	168
INDEX	171

ILLUSTRATIONS.

THE MAMMOTH	*Frontispiece*
I. UPHEAVAL : SUBSIDENCE : DENUDATION *to face page*	51
II. DIFFERENT KINDS OF PLANTS OF THE COAL FORESTS . . . . . ,,	65
III. TRILOBITE . . . . ,,	79
IV. FOOTPRINTS OF LABYRINTHODON : FOOTPRINTS OF BIRDS, (2) WITH MARKS OF RAIN-DROPS . . . . ,,	83
V. FISH-REPTILES . . . . . . ,,	87
VI. BIRD-REPTILES . . . . .· ,,	93
VII. FOSSILS OF THE CHALK . . . . ,,	97
VIII. GIGANTIC IRISH STAG (CERVUS MEGACEROS) . . . . . . ,,	108
IX. THE MEGATHERIUM . . . . . ,,	112
X. 1. FLINT ARROW-HEAD ; 2. STONE AXE IN HANDLE ; 3. FLINT KNIFE ; 4. BONE HARPOON ; 5. BONE NEEDLES ; 6. SCEPTRE MADE OF HORN ; 7. MARROW SPOON . . . . ,,	129
XI. EXAMPLES OF PRE-HISTORIC DRAWINGS . ,,	135
XII. LAKE-DWELLINGS . . . . . ,,	148
XIII. THE GUADALOUPE HUMAN FOSSIL . . ,,	159

THE
PUZZLE OF LIFE.

THE FRAMEWORK OF THE PUZZLE.

You must often have looked with wondering eyes at this World of ours, and asked yourselves questions about it. How did it come here? What is it made of? How old is it? All of them questions not to be answered without a great deal of thought and study, and even then not so perfectly as we should like. It is easy to say "It is here," and "It is made of earth," and "It surely must be old," but that will not satisfy us. We want to know something more certain than this, if possible. We can see that a clock goes with wheels, but we are not very intelligent people if we do not want to find out what makes the clock go. One way of finding out is to pull things to pieces, but we cannot exactly do this with the World. We must think about it, and put together all the knowledge we can gain from the outside and inside, and from the other Worlds around us, which we can see, and when we have done this we may get something like answers to our questions.

How did it come here? But this is not quite the right way of asking the question, because the World is never for two moments together in the same place. It is travelling in a great circle round the Sun at the rate of more than sixty thousand miles an hour, and has been ever since it was formed. That is a wonderful arrangement by which all Worlds travel round some other World larger than themselves, in greater or less circles, and we do not know why it is, though we are certain that it is so. The Moon travels round us once in about every month, and we and the Moon together round the Sun once in every year.

Then again, other planets, with their

moons, such as Jupiter, for instance, travel round the Sun in much larger circles than our World, and take many years to do the journey, while Venus, which is nearer the Sun than we are, travels in a much smaller circle, and takes less time. We do not perceive that we are moving so fast because everything we see is moving equally fast with us; but there is no doubt that we are spinning along at sixty thousand miles an hour.

If we ask an astronomer how our World came into existence, he will tell us that it is probably a mass separated from the Sun, that it was once red-hot, and that it slowly cooled down until animals and plants could live upon it. He will tell us besides, that he can see mountains and valleys in our Moon, and land and sea, snow and clouds, on the planet Mars, with his great telescopes. When he thinks about the planets and our own World, then he believes them to be pieces of some much larger World—perhaps the Sun—which now travel round the Sun and receive their light and heat from it. The World is made of what we call "earth," and it is of this I mean to tell you now—how it was formed, what changes have taken place in it, what plants and animals have lived upon it, and what reasons there are for thinking that it is an exceedingly old place, with a long and interesting story to tell.

Little was known thirty or forty years ago by the most learned men about the age of our World, and it was thought that the human race had not lived here very long. It was indeed known that many large animals, whose huge bones have been found, must have lived before man came to inhabit the Earth, and that even far smaller creatures—such as fishes, and crabs, and insects, and shell-fish—most probably lived for many generations, and died and left their bones and shells in the soil long before the first man or the first tribes of men came to share the World with them. I hope to be able to tell you something of the strange and beautiful history of all these animals, and of man himself, and to show you what reasons there are now for thinking that the human race has inhabited this Earth for a very long time indeed, and how all this knowledge has been gained and put together piece by piece. It is something like the different parts of a puzzle-map, which might be scattered all over the house, and found at one time or another in different places, and at last made up altogether. Some parts of the puzzle have not been found yet certainly; but so many have been collected, and they fit into one another so well, that we can begin to see its real shape and size. It will perhaps be a very long time before some of the missing pieces are found; but in the meantime we can go on without them, and put the framework together, and no doubt in time we shall see what our puzzle, the history of life on the Earth, was like.

Before telling you what its parts are, I ought to say where many of them have been found, and how they are still being looked for. They are found *upon* the ground, *under* it, in caves, in rivers, and in the sea. Since railways have been in use a great many tunnels have been made, as well as very deep cuttings through hills, and some of these are several miles long. In this way we have come to know something of the Earth below the surface. Some of these tunnels are bored right through high hills and even mountains, and the cuttings are deep enough to hide high houses if they were put into them. While digging these the workmen have found many of the parts of our puzzle, which are the bones of animals, and fishes, and shells, and even smaller things—such as insects. These could not possibly have been put there by anyone, because they were many, many yards below the surface, and, until they were dug up, nobody imagined that they could be there. Many other things besides have been dug out of these places, but nearer the surface, such as weapons and tools made of flint, and stone, and bone, and metal, and pieces of rough crockery, and various ornaments, all of which must at some time or other have been made and used by people very like ourselves. In digging canals, too, the same kinds of things have been found, and some caves are almost filled up with them. We have other means, too, of knowing what is under the surface of the ground we walk upon. Many of the coal-mines are so deep that the Tower of London, or St. Paul's Cathedral, or York Minster, or even the Pyramids of Egypt could be buried in them! In digging these the workmen have had to go through a great quantity of earth, sometimes chalk, sand, or gravel, or clay or limestone, layer upon layer, placed, like a pile of books of different kinds and different thicknesses, one upon the other, until they have come to the coal. In these different layers of earth parts of the puzzle have been found, and we shall see by-and-by what parts have been found in the coal itself. Then again, when deep mines are made to get the metals, iron and gold and silver, these layers of earth have to be dug through; and when the beautiful kinds of stone, like marble and limestone, are wanted, they must be dug out of the sides of the hills, and in doing this still more pieces of the puzzle come to hand. But there are other places where Nature herself seems to have shown us some of them without the trouble of searching for them. In many parts of the World, by the sea, and on the banks of rivers, there are cliffs hundreds of feet high, like the chalk cliffs at Dover and Ramsgate, and the sandy cliffs at Folkestone and on the south coast of Devonshire. These cliffs have been cut into by the sea very gradually, and a kind of wall has been left, and from the sides of the cliffs great numbers of the pieces of the puzzle, bones, shells, &c., have been collected and taken away to museums. But the little we can do with our mines and railway tunnels is nothing in comparison with the work of Nature. In some of the great mountain chains—the Andes, the Himalayas, and the Alps, for instance—parts of the sides of mountains have fallen down, and rents many miles long have been left, showing what had been buried there in the different kinds of soil; and where rivers have cut deep, narrow channels through the earth, like the Cañons of Colorado, these natural miners have turned out more of the parts of "the puzzle of life" than we can with all our labour.

It will not be easy at first to understand all the wonders I have to show you, but, when we get further on, you will see them one by one, and there will be very little difficulty. You know now where these things are to be found: principally in the ground you walk upon, without knowing all there is beneath you. The creatures here are much more wonderful than any of the monsters of fairy tale or fable, because the works of God are greater than the imagination of men who have invented the stories of flying dragons and griffins, and trees which grew up into the skies; but I cannot help thinking that this imagination shows what men thought *might* once have been, and we shall see that "truth is stranger than fiction." Creatures really did live on this Earth of such strange shapes and great size that the imaginations of those who wrote the fairy tales did not exaggerate much; and, though we know that no flying serpents or immense birds like the Roc are living now, and that there is no beanstalk which grows up into the sky while we are asleep, we shall see that there were lizards as large as whales, and birds taller than elephants, and great sloths stronger than the rhinoceros or hippopotamus, and ferns as high as oak trees, and mosses as large as gooseberry bushes; and that perhaps these animals and plants grew much faster than they do now, and that their dead bodies form a very large part of the earth of our World. This is not imagination, and when you go to a museum you can see all these wonders for yourselves, just as they were taken out of the earth; but of course the bones only of the animals are there. The flesh has long since gone away, and some of the stalks and fronds (leaves) only of the ferns remain to show us how large they must have been when they were alive and growing.

It will be necessary to use a few scientific names, most of which are borrowed from the Greek and Latin languages, but I will explain the meaning of them all, so that they will be easily remembered. First of all, then, the pieces of the puzzle are called *fossils*, and the name comes from a Latin word meaning "dug out;" because they have been dug out of the ground either by man in making railways and mines, or by Nature in the many ways in which she works by cutting down cliffs and scooping out valleys. These fossils are bones of animals and fishes, the skins, shells, and wings of insects, and the stalks and leaves of plants, some of which have lain so very long in the ground that they have become as hard and heavy as stone. But the shape of them always remains, and the moment you look at them you see that they once belonged to living creatures.

I shall give you pictures of some of these fossils ; and no doubt you will be able to find some like them in the chalk and sands of the seaside—beautiful shells and bones of fishes. You may pick these out of the cliffs, and then go to the pools of salt water left among the rocks by the ebbing tide, and compare your fossils with the living shell-fish, and see how nearly those inhabitants of the ancient oceans resemble the creatures we find now, sporting in the water, just as these fossils did when the sand and chalk cliffs were under the sea. Of course all the bright colours are gone from the fossils, for the colour of animals fades away soon after they die, and the flesh does not last long; but the hard parts—the bones and shells—are not easily destroyed, because they are made of the same material as rocks. And when we look at the fossil plants we see the same thing. The colours of the green stems and leaves have quite faded, but the delicate shapes of the leaves and branches, and the grain of the wood, can still be seen, and you will have no doubt that they once lived and bore flowers and fruit, and died, as plants are living and dying every day.

You have got so far now that you know what fossils are, and where they may be found. You know that they are the small and large pieces of the " puzzle of life "—of all sorts of different shapes and sizes—and you know that they are scattered about the Earth, deep down in coal-mines, on the tops of mountains, at the bottoms of rivers, in deep caves, and under the sea. The patience and industry of clever men have been well spent in gathering together all they can find, and arranging them in museums for our instruction, and making a history of them which is more wonderful than the Arabian Nights, and more beautiful because it is all *true*. And, though you may think it strange that I promise to show you creatures more marvellous than those of the fairy tales, I shall keep that promise faithfully. We shall find no Genii with wonderful lamps and magic rings, because they never really lived, though it gave us much pleasure and amusement to read about them; but we shall see what God, the greatest Genius of all, has done by means of His magicians — the laws of Nature. These magicians have built up high mountains and dug out valleys, and sent mighty rivers sweeping down to the sea, and even filled up oceans with sand and chalk, and buried ancient forests deep down under sea and land. They worked with fire, and air, and water; not quickly, but with such strength that nothing could resist them, and they gradually moulded the Earth into the beautiful thing it is, so that

>In contemplation of created things,
>By steps we may ascend to God.—*Milton.*

But, lovely as the Earth is, we should not perhaps have thought so much of it if there had been nothing to discover. We see that it has been prepared for us an immensely long time ago; and when we know a little, we want to search further and find out what the whole plan of Creation is, so far as we can. You will be surprised when you know how many signs of past life there are around you—many more than you can see with the eye. The Earth is one great burying-place of creatures which have passed away. You are walking over their dead and fossil bodies at almost every step. They are built into the walls of our houses, and there are millions of them in some of the commonest stones of the pavement. Those round, smooth pebbles, called flint stones, which we pick out of the gravel walks, were once partly such soft tender things as sponges; but time has hardened them, and they have been rolled together in seas and rivers by the always moving water until they have become quite different to look at from the rough blue flints they were when they were washed out of the chalk beds. When you are walking along the sands of some seacoasts, you are treading on little specks of these small flints which have been ground down fine in that great mill, the ocean. The sponges, then, did some part in the building up of the Earth. The very chalk you draw with is composed of the shells of sea-animals. Your slates and slate pencils were once a fine mud at the bottom of the sea, since become so hard that it is used for covering the roofs of our houses, and in this mud lived myriads of small shell-fish which have sometimes left their frail houses in the slate beds to tell us how they were made. That slate is the hardened mud of an old sea bottom, there is no doubt at all.

There are many other things in common use which show us the life that was.

Perhaps you did not know that coals are *compressed plants*, and that we are now burning the vegetation of the past time! But these will be described in their right places by-and-by, and you will see how certain it is that some of the commonest things we use were living creatures and graceful plants.

Here is "the framework" of the puzzle, and I think you will agree with me that we shall have pleasure in putting it together with all the queerly-shaped pieces we shall find in the following chapters. We have fossil plants to show us what grew upon the Earth, fossil bones to tell us what animals lived here, and thousands of different kinds of fossil shells and fishes to show us that the seas in the long past time were crowded with life; and besides, though there are no written histories of the men whom we shall read about, they, too, have left many things which they used in the caves where they lived and in their graves, to make us feel certain that they were some of the oldest people that ever lived. With all these things to help us, it will be strange if we cannot make out a great deal of the history of life upon our Earth.

THE GEOLOGICAL PART.

You will have learned from other books something about the size and shape of our World: for instance, that it is a great round body, or rather more like an orange, a little flatter one way than the other, and about 8,000 miles through, from one side to the other, and that it turns round once in every twenty-four hours; but I have only to tell you now what it is made of. The material is called rock, earth, or soil; and there are many kinds of it, such as granite, gravel, clay, sand, chalk, mud, and so on; and we shall see that many of these different soils contain different fossils.

It is supposed that a very long time passed while these were being laid one upon another, and before many plants or animals lived here, and there are good reasons for thinking that underneath these soils the Earth is very hot, perhaps in a melting state, because we know that volcanoes like Vesuvius and Ætna throw out flame and smoke and lava, which is melted earth and rock; and that this lava has run down the sides of the mountains for miles, in a great stream of liquid material, and covered up and destroyed whole villages and towns. You have heard of earthquakes, when the ground shakes and cracks, and houses are thrown down, as they have often been in Spain, Italy, and South America. This convinces us that the inside of the earth must be very different from the outside. Two or three years ago Mount Vesuvius was boiling up, and the people of Naples feared that it would throw out some of the terrible lava and red-hot cinders, and burn up their vineyards and perhaps injure their city; and during the last two or three years many people have been killed by earthquakes in South America. These things seldom happen in the North of Europe, and when they do they are only slightly felt, and people are not killed, neither are houses thrown down.

Still, this shows that there must be some great force underneath us, and very much heat. We see nothing of this when we look upon the green fields, and we should scarcely think it possible if there were not histories about these eruptions, as they are called. But when I tell you that I have felt the Earth tremble, and seen fire rushing out from the top of a high mountain whose sides were covered with snow, you will understand how real it is—though it may seem so strange.

People at one time liked to fancy that powerful spirits lived in volcanoes and made them their workshops: but we know better now.

Well, the interior of the Earth is evidently very different from the part we live upon; and it is the outside we have to think about now, which would be dreadfully cold if the sun did not shine upon it, though the inside is so hot.

I have called this "the Geological Part," and the name Geological comes from two Greek words meaning "a talk about the earth;" but now you know it in its English dress it will be easy to recollect it. Geology is then the study of the many kinds of rocks and fossils which make up our World, but we must know something of the way in which they are placed.

You may have noticed, if you have made many journeys to different parts of England or Wales, that the rocks or soils are very different in various places. Sometimes we find numerous chalk-pits, as in parts of Kent, or Sussex; if we go into Devonshire we may notice the very red colour of the soil and of the cliffs, especially near Sidmouth, Dawlish, and Teignmouth; in North Wales we find great quarries and hills of slate; while around London we see a great deal of clay used for making bricks, and called the London clay, as well as many pits in gravel so useful for making paths and mending roads, and in Kent and Sussex chalk cliffs and hills are common.

Now after studying these various rocks all over our country, we find that there is a certain regular order in which they are found; some have been made a long time before others, and while most kinds contain some fossils, those found in the oldest rocks are much less like the living plants and animals than the fossils we find in the newer rocks.

But you will want to know how it is that we can tell that one rock is older than another, when both appear at the surface of the earth. It would take a long time to make sure of this for ourselves, but it will be enough to say that the various cliffs, quarries, and railway-cuttings often show one kind of rock resting upon another, and these always occur in a certain order. Thus we never find the Chalk resting on the London Clay, but we constantly find the London Clay resting on the Chalk. And this is proved in another way, by deep well-borings. Underneath London many wells have been carried down right through the London Clay, and if only continued deep enough they always reach the Chalk. In the same way, the order of the other rocks has been ascertained in different parts of the country, by examining all the pits and quarries, and cliffs and cuttings, with the help of what knowledge can be obtained from deep mines and wells.

You will now begin to wonder why the

older rocks should appear at the surface. I have told you about earthquakes, and you will find that many dreadful earthquakes must in former times have ravaged our country. The reason why the old rocks come to the surface is because they have been lifted up sometimes violently, but more often very slowly. And the newer rocks which formerly rested on them have very often been quite washed away, either by the sea or by rivers and little streams which formerly acted upon them.

Suppose then we take six books, some thick and some thin, and pile them up together on the table, the lowest being a good thick one. The lowest we will call granite, the next slate, the third sandstone, the fourth coal, the fifth chalk, and the sixth the London clay. These will represent some of the principal kinds of earths, and you can fancy many more with other names coming between them; but the London clay can never be below the granite nor the chalk below the coal, for the great coal beds were formed long before the chalk and clay. They generally come in much the same order as we have named them, hard rocks like granite at the bottom, and softer earths, like sandstone, chalk and clay, a long way above them. But we do not always find all these earths in one place even if we dig ever so deeply, though the granite would always be found at the bottom if we went deep enough.

Sometimes the granite and other old rocks have been pushed through the upper layers by some great force, and have broken them and risen above them in magnificent mountain chains, like those of the Andes in South, and the Rocky Mountains in North America, the Wicklow Mountains in Ireland, the Grampians in Scotland, and the Cornish mountains in England. We can easily suppose that the lowest of our books (the granite book) has been pushed upwards by some great force from below, and parts of it broken through the others, and raised high above them; and this is what has actually been done with real rocks. And as this kind of upheaval has taken place at different periods of the earth's history, we find that granites have come to the surface at different times.

When the layers are thus broken through they are often tilted up on end and tumbled about in confusion. But where there has been no disturbance like this, they generally rest evenly upon one another in their proper order.

Granite, and rocks of the same kind, are not in the least like chalk, or clay, or even sandstone, and when once you have seen any of these you will not be likely to mistake it for the others. Granite is excessively hard, and has a beautiful appearance when polished, with a number of brilliant white and some dark specks in it. It is used for paving the streets of towns, for which purpose it is cut into oblong blocks, and for the pillars of fine buildings. Sometimes it is dark brown, sometimes reddish, but generally a bluish grey. This rock is composed of a great quantity of crystals, and for this reason it is thought it must have been melted at one time by intense heat in the earth, and afterwards slowly cooled. Chalk is very different, and sandstone, though it is also hard, not in the least like granite.

HOW THE ROCKS WERE FORMED.

What I have just said is about all that we know of the formation of the oldest and hardest granite rocks : but there is something going on now which confirms the belief that the materials of which they are made were melted together by a greater heat than we can make in our furnaces for melting iron ; for I should tell you that it is easier to melt iron and copper than granite rocks. Volcanoes often throw out melted earths which when cooled appear to be made of much the same materials as these granites.

SANDSTONE.

But we know more of the manner of the formation of sandstone. This rock is composed of rounded grains of sand just like that we find upon the sea shore. If you take a handful of this sand and squeeze it tightly, it will keep together a little while. Now suppose a quantity of this sand was pressed by a very great weight—the weight of a large hill for instance—after many years the grains would stick firmly together, and become a sort of stone. It is in this way the sandstones must have been formed, and perhaps heat helped the work, though not so great a heat as melted the granite. The sand, after it had been washed upon the sea shore, became gradually covered with other earths hundreds of feet thick, and the immense weight above it pressed it into stone : but you may imagine how very long a time it took to do this. Sandstones are used for building, but they do not last very long ; the frost makes little cracks in them and they soon crumble away to the grains of sand of which they were made. Several fossils are found in some of these sandstones, which have been formed at many different periods of the earth's history.

CHALK.

You have seen those high cliffs of chalk along the south coast of England, perhaps, and you have wondered what that beautiful white earth was, and how it came there. It is found in many parts of the world, and the south and south-east of England are to a great extent composed of it. The material is called by chemists carbonate of lime. It is almost entirely made up of minute shells called *foraminifera*, from two Latin words which mean that there are many openings or chambers in their shells, and there are many beautiful fossils called *ammonites* imbedded in the chalk. These are shell-fish, two or three inches, and sometimes a foot across, and their shape is very like that of the young leaves of the common fern before it has opened in the spring.

Millions of these tiny foraminifera are living now in parts of the Atlantic and Pacific oceans, and when they die their shells sink to the bottom and form a greyish mud, something like chalk.

When H.M.S. "Challenger" was sent out in the year 1873, to find out what was at the bottom of the deepest seas of the World, great interest was felt in the expedition, because we were sure that we should learn something about the manner in which some of the rocks were formed.

We knew that the whole of the beds of the present seas must be receiving the washings of the rivers and the bodies of many fishes and animals, and that the rocks of the future must be forming down there by these accumulations. Long lines were let down from the ship with a dredge at the end, and thus parts of the bottom of the sea were brought up and carefully examined. It was found that the washings, stones, clay, and mud of the land were carried hundreds of miles out to sea, and laid upon the bottom. But in the deeper parts, where the Alps would be almost covered—there was a fine grey mud composed almost entirely of the shells of the little foraminifera, and this, no doubt, is the chalk of future times, or perhaps limestone of a harder kind. Deeper, too, than where this grey mud is found, there is a reddish mud, exceedingly fine and soft. We cannot exactly say yet whether this is formed from the remains of shell-fish; but it is, at all events, very like the clay of the land, and in some future time will most likely become like that stiff mud we know so well. So that even the materials for bricks are being made now, and perhaps when all those hundreds of islands scattered about the Pacific Ocean are joined into one great Continent, this red mud will be raised and made use of for building the houses of new peoples and nations.

When we see this going on now, of course it is very easy to conclude that the chalk, a great deal of which is above the sea now, must have been formed in the same way at the bottom of an ancient ocean, and afterwards raised by the same kind of upward force which made the granite break through other earths.

If we did not know that the same cause was at work now, and that the same kinds of shellfish were living and laying down new beds of chalk under the sea, we should not know how to account for the quantities of chalk in the world. For innumerable agest hese little creatures have thus been paving the floor of the ocean with their dead bodies, and you may suppose that countless millions of them must have lived and died! In some of the chalks the shells of the foraminifera can be quite distinctly seen with a microscope, and when these are compared with the shells of living ones, they are seen to be almost exactly alike. Next time you pass through one of the railway cuttings through the chalk in going to Brighton, or Ramsgate, or Dover, remember that those high cliffs were built up by these Liliputian giants under the sea, and you may think of the chalk as "foraminifera earth."

COAL.

You see this black shining substance almost every day, and you know it is dug up from very deep pits where the poor miners are often killed by explosions of gas escaping from it. But it is as well to know what it is and how it comes to be so useful to us. In the language of chemistry it is called "carbon," and a great writer has given it the poetical name of "compressed sunlight." But you will ask how sunlight could possibly get into a deep mine, and how it could be compressed there. You will see that the explanation is really quite simple by-and-by. This coal was once above ground, and was a splendid forest of waving palm-trees, and ferns, and gigantic mosses, as you will see by the pictures of the fossils of them.

Many of the animals and plants of past times were giants compared to those living now, of the same species or kind, and many of the plants of the present time are dwarfs to those of the same kind which formed the coal beds. Many generations of trees must have grown and died, and others must have sprung up, and so on, until beds of them, some ten, others twenty, or even thirty feet thick, were formed. Here, buried in the coal, are the stems, leaves, bark, roots, fruit, and seeds of these trees, and we can have no doubt that almost the whole of the coal is composed of them. You must not expect to find the shapes of these in every piece of coal you may happen to look at, because most of it has been greatly changed by the great weight and pressure upon it, and the length of time : but it is certainly all the same substance—wood turned into coal. The fossil plants of the coal are of course entirely black, but there is no mistake about their having once been living plants.

You will ask perhaps how the coal came to be buried so deep. It is not so always, being sometimes at the surface. But just as the granite has been pushed up through the other rocks, so has the coal in some places been uplifted and in others has sunk down. It was often covered up by other earths to a great depth, after the trees which composed it had died; but where it is now at the surface these newer earths have been afterwards worn away. When the sun shone upon these coal trees they took its warmth and light into their stems and leaves, for they could not live without, and this made them grow so fast and become so large that it is not untrue to call coal "compressed sunlight." Charcoal is in some respects so like coal that it would seem to you at once that they were probably the same material. Charcoal is simply burnt wood, and when the coal forests had died down, and when these beds sank down beneath other layers the pressure and heat together turned the wood and leaves into a hard mass like charcoal in colour, but heavier and more solid, and just enough of the stems and leaves have been left to enable us to know with certainty that coal was once wood.

We light our fires now and drive our steam-engines with the heat of the sun which shone upon the coal forests, and has been stored up for many thousands of years in the Earth, to be brought out once more to give us light and warmth.

CLAY AND MUD.

While the ancient forests were growing up to form the coal beds, and the foraminifera were slowly building up the chalk, as I have explained, the Earth was covered with water in some places which are now dry land, and the sea now flows over parts of the World which were once the habitations of plants and animals. These great changes have left their marks upon many a mountain side, and many an old river or sea bed has become filled up. A map of Europe during the chalk period would show that the places where Paris, London, Copenhagen, and Berlin now are were then under the ocean; but since then these places have been lifted up, and mud, clay, and gravel swept over the chalk in many places by the action of new rivers and seas. Water, you perceive, has had a great deal to do with these changes, and indeed it is one of Nature's most powerful tools, for it can wash down rocks and cliffs and cut its way in rivers for thousands of miles over the Earth's surface. It carries down mud, and clay, and gravel, and this soil, which has been named alluvium, is one of the most interesting of all to us, because it contains the bones of the immense animals we shall talk about presently, as well as those of the oldest races of men with their weapons and ornaments.

The mud age, and we are in the mud and gravel age now, belongs to what is called the Tertiary period, and we shall see that this age has lasted a very long time already, so long that though it is still going on, the most extraordinary animals have lived and died, and not one of them is now left alive. Still the same washing and cutting of water is going on which buried their bones in swamps, and bogs, and river caves, and may perhaps carry away some of the bones of us who are living now, to be found ages afterwards by future generations who will read our history in these silent witnesses, as we read the history of the tree fern sand foraminifera in the coal and the chalk.

The present age of the World's history is the Mud age, or, as we shall call it in future, the Tertiary period, and I think you will agree with me when I come to describe it, that it contains the most interesting of all the pieces of "the puzzle of life."

The earth of the Tertiary period is very different from a great many of the older earths. Clay, mud, and gravel are the washings only of the older rocks, the fine particles which have been worn off from them by frost and water and carried down by rivers and left in large beds, and sometimes they have a good deal of decayed wood and weeds mixed with them. Here are found the bones of the great animals which were so much larger and stronger than those of the same kind living now, or any that lived before them.

UPHEAVAL AND DEPRESSION.

These two words are so often used in books on geology that we shall not be able to get on without knowing their meaning. We have seen that the rocks have been formed in a certain way—some by heat, some by water, and some by dead forests — and that they lie over one another in pretty regular order. But this order has sometimes been disturbed and the layers have been tumbled about among one another very much. In some places the older rocks, such as granite, slate, and sandstone, have been pushed up through those above, and in others the coal has sunk down and been covered with thick layers of chalk, sand, and mud. When the force below pushes a layer up through the others it is called *upheaval*, and when a layer sinks down it is called *depression*, or *subsidence*. Both these actions are going on now in different parts of the Earth. A great part of Sweden, Norway, and Denmark, of Spitzbergen, Siberia, and the north of America, is being slowly raised higher above the sea, as we know by the height their old sea beaches now are above the water; while part of the shore of America opposite to Europe and also the south of Greenland is slowly sinking down, as we know by the remains of land animals and trees which are now covered by the tide; and at many places on the coast of India this subsidence is also going on. Nearer home, too, there is an example of it in the island of Guernsey. All round the coast of this island, like that of Jersey, are found tree trunks and other remains of old forest land beneath the water. Old histories refer to this as dry land; and if a map of it made in 1406 is correct, this land must have sunk about 150 feet since that time.

Thus we can see, even at the present time, the very same changes which have worked upon our Earth for innumerable ages. It is now easy to understand how the forests which must have grown above in the air have, after a long time, sunk down to a great depth, and been turned into coal, and covered with the sediment, sand, gravel, and chalk from the seas which afterwards flowed over the places where they grew.

Sometimes the rocks by the sea shore are cut into terraces or steps by the constant wear of the water, and when we see these water marks far above the present level of the sea we know that the land must have been lifted up gradually above the sea. There are many such terraces in Norway. To prove whether this is so marks have been cut upon rocks at a measured height above the sea, and after some years these marks have been noticed to have been raised much above the water by the "upheaval" of the earth at that place.

Generally this change of level has taken place gradually, and the greatest work in moving the layers of earth and displacing them has been very slow. But in some places violent and sudden shocks have happened, tearing up the rocks and piling them up in heaps; and now and then islands have suddenly appeared in the sea and vanished out of sight completely in a short time. Islands have thus come up in the Mediterranean Sea within the memory of man. In the year 1831 the island of Julia suddenly appeared near the coast of Sicily, and since the year 186 B.C. no less than three islands have started up in the bay of the island of Santorin. In this century islands have appeared among the Azores, the Indian Archipelago, the Philippines, the Moluccas, and on the coast of Kamtschatka and other places. Some of these have appeared suddenly, others slowly, and they no doubt have been raised by a great force from below.

You will see now how easy it is to account for the changes of the places of the layers of rock. The same thing is going on now which has been going on throughout all time, only perhaps with more energy formerly than now, making mountains, islands, and continents, raising up a large tract of land in one place and sinking an island or a sea shore in another.

These changes have been of great use to us too. Suppose all England had been covered with coal or slate, we should not have been able to grow anything! As it is we have sand and gravel in one county, chalk in another, slate or granite in another, and coal down below in several, and we can grow a great variety of plants on all these different soils. We have to thank "upheaval" and "depression" for this. The force which is always working below us has turned up the different soils like a gigantic plough, and brought some to the top and covered others, so that instead of having to dig down deeper than ever we have yet, we have only to go from one county to another to find the different rocks. We know that we could not get at the coal in Sussex without going down an unknown depth through the chalk and other earths, but we dig for it in the North of England, where we know its depth below the surface.

I will try now to give you some idea of the way in which the rocks come in their order, or the succession of formations as geologists call it. If we started to walk from Wales to London the rocks we should pass over would be — slate and flagstones in Wales, and going on towards London, limestone, old red sandstone, more limestones, coal beds, new red sandstone, oolite, greensand, chalk, and last London clay. We might not always find each of these near the surface, but they would be found to be the principal rocks on a line between Wales and London, the oldest being in Wales and the newest or most recent as we get nearer London. That word "oolite" which I used comes from two Greek words meaning "roe" and "stone," because the rock is composed of little rounded grains of a chalky substance shaped like the hard roe of a fish, or like sago before it is cooked.

If you look at the following table you will see how the principal rocks are placed one upon the other, beginning at the lowest or oldest at the bottom and going up to the newest at the top of the table, and on the right hand side I have written the names of the principal fossils which each kind of earth contains.

TABLE OF THE SUCCESSION OF FORMATIONS.

TERTIARY, or Upper Rocks		Peat-bogs and caves River-mud and brick-earth, gravels, and boulder clay (alluvium)	Fossil Man, with stone implements, &c., mammoth, hippopotamus, rhinoceros, Irish stag, cave lion, &c.
		Crag of Eastern Counties	Numerous shell-fish, mastodon
		London clay, &c.	Turtles, crocodiles, shell-fish
SECONDARY, or Middle Rocks	Cretaceous	Chalk (with and without flints) Greensand and gault Wealden clay, &c.	Foraminifera, &c., sponges, corals, sea-urchins, shell-fish (Belemnites, Ammonites, &c.), fishes
	Oolites	Portland stone Kimmeridge clay Coral rag Oxford clay Cornbrash and forest marble Great oolite Fullers' earth Lower oolite	Immense reptiles, the Ichthyosaurus, Plesiosaurus, Megalosaurus, Pterodactyl, &c. Animals allied to the opossum and kangaroo
		Lias clay and limestone	Cycads and other plants
		New red marl and sandstone	
PRIMARY, or Lower Rocks		Coal Millstone grit Mountain limestone	Ferns, club-mosses, a few firs, calamites, &c., in great abundance
		Old red sandstone Silurian limestones and slates	Numerous corals, shell-fish, trilobites, fishes, &c.
		Cambrian slates Laurentian rocks	The Laurentian rocks contain the oldest known fossil, the Eozöon (or "life-dawn animal")
IGNEOUS, or Volcanic Rocks		Greenstone, basalt Porphyry Granite, &c.	Of various ages (no fossils)

If you read this table upwards from the bottom you will notice that life began in a very small way with Eozöon (the "life-dawn animal"), that fishes appeared afterwards, that the wonderful forests of the coal period then grew and were covered up by other rocks and pressed into solid coal, that numbers of great crocodile-like animals lived all through the oolite time, how the deep wide beds of chalk were laid down by humble foraminifera, and when we get to the recent newest beds of gravel, mud, sand, clay, &c., the sweepings by water of the older rocks ground down by ages of wear and tear, we have the mammoth, mastodon, megatherium, and other great vegetable eaters, and lastly Man himself with his simple weapons of stone, bone, and horn—our early forefathers.

You must always keep in mind that the greatest of these changes have taken place very slowly. Mountains have been raised, and whole continents have been sunk by movements so slow that if the hands of a clock went only once round the dial in a year the hand would go faster than these mountains have risen or the continents sunk. Almost always whenever there has been sudden and *violent* action it has been near volcanoes or during earthquakes; but these things, terrible as they are to the people living near, disturb only a very small part of the surface, and such violence neither buried the coal beds nor raised the slate hills of Wales. Many of the small effects of the internal force of the earth have been sudden and violent, but the greatest and grandest have been slower than anything we can imagine.

If this had not been so, we should not find fossil shells just as they sank quietly to the bottom of ancient seas, quite undisturbed. We should not find delicate ferns and insects with all their breakable parts perfectly preserved, and as lightly laid as if you had put them away carefully in a cabinet upon cotton wool. Yet many of these have sunk down hundreds of feet below the open air where they *must* have lived. We find the ripple marks of the waves on old sandstones, and even the prints of the feet of birds and animals as they walked upon that rock when it was soft sand, and the little pits made by rain-drops on the moist earth. All this speaks of stillness, and gentle movement, no violence. So slowly and softly have these rocks settled down, that we can read in them the history of the life that was. But if there had been any sudden and rough movement all these fossils might have been broken up and we should have had nothing but fragments, and the "puzzle of life" could never have been put together. Nature's forces are immense, but they work slowly, irresistibly, and majestically.

THE ICE AGE.

We have seen now what the principal rocks are made of and the way in which their places have been changed by upheaval and depression. Water, as we know, has been at work and has done great things in *all* ages of the World's history. I have called it "one of Nature's most powerful tools," and when we look at the quantity of chalk alone that there is in the world, and remember that this was all laid down in water, and perhaps a great part of its lime carried down by rivers to the seas where it settled to the bottom, after the corals and small shell-fish had worked it into their bodies, we are right in thinking water a great Magician indeed. Why, even so small a river as the Thames carries down to the sea every year as much dissolved earth as would make a good large hill; and what must such rivers as the Nile, the Amazon, the Mississippi, and the great Chinese rivers do! There must have been gigantic rivers, too, in the old times, or else it would have been impossible that the deep sandstone and slate beds could have been formed; for these are all laid down by the washing away of earth in water.

Ice, which is only solid water, has also been a powerful tool in shaping the surface of the Earth, but it has not been *always* at work as water has. Ice now covers only a comparatively small part of the globe near the north and south poles, and mountains like those in Switzerland; but by watching what ice is doing now in these places we are able to be certain that there has been a time when it covered Scotland, Cumberland, Wales, Sweden and Norway, and nearly all North America. In watching the great "rivers of ice," called glaciers, in the Alps, for instance, we see that they slip down from the mountains slowly, creeping on year by year, and bringing with them pieces of rock and stones. We see also where they have melted that they have been grinding the rocks beneath them with their great weight, and have cut grooves into, and scraped and polished the hardest granite. The stones underneath the glaciers have been pressed so heavily upon the rocks that they have left deep marks, and we find the same kinds of marks and heaps of stones in many mountains where there are no glaciers now. There are other things too which convince us that a great ice sheet spread over almost the whole of Great Britain. When the huge icebergs break away from the frozen shores of Greenland and North America, they often have frozen into their ice large blocks of rock, sand, gravel, &c., and when they drift into the warmer seas of the south they melt, and of course these blocks or "boulders," as they are called, sink to the bottom. Just the same kind of boulders are found in many parts of the world, where icebergs never come now, and as they are of a different rock from that on which they lie, they must have been brought there somehow. We naturally suppose then that they were brought by icebergs. Sometimes boulders of granite have been found thus among clay, many miles from where there are any granite rocks on the surface, and there can be no doubt that they were originally frozen into an iceberg, which floated away with them and when it melted left them so far from their native place. In many of the midland and eastern counties once floated these icebergs, dropping the stones and boulders which they brought away from the Welsh, Cumberland, and Scottish mountains.

The climate of the earth must have been fearfully cold when our country was covered with ice, just as Greenland is now. Geologists suppose that there must have been more than one age of ice, and that between these ages the climate of the world was pretty much the same as at present, although it is certain that there were periods when England was much warmer, because many of the fossil plants could not have grown in a cold climate.

You will want to know whether there were any land animals living during the ice periods. It is impossible to be quite certain, but it is most likely that the mammoth was living both before and during the *last* ice age, because its bones have been found among the earths brought down by the glaciers.

I have said all you will be likely to remember at present about the nature of the different rocks, but it will help you to understand better how they have been laid one upon the other, and how they have been moved and broken by upheaval and subsidence, if you look at the drawings on page 51.

DENUDATION.

It has often happened that some of the harder and older rocks, like granite and slate, have pushed themselves through those earths lying above them, and then the sea or a great river has washed away all the earths from one side of the rock. The rain, too, falling for thousands of years, has swept them down into the valleys and mixed them together. This is called denudation, or "laying bare" the harder rocks by washing the softer ones away from them. Those beds of pebbles on the sea shore also have been battering against the rocks for ages and very gradually wearing them away, as you can see if you watch the stones being driven into and sucked out of holes and cracks by every wave. Thus, both the loose stones and the solid rocks get polished and ground away, and Nature is always destroying and making again by turns. If this destruction went on continually without any raising of the land to make up for it, the surface of the whole Earth would in time become level; but old sea beds are always being slowly raised above the water and prepared for the growth of plants and the habitation of animals.

If you watch the little rills of water on

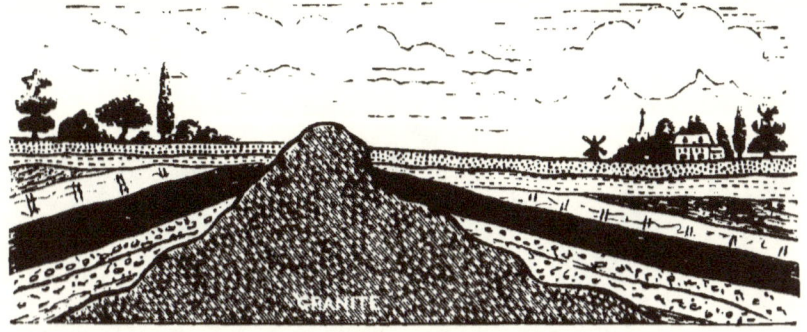

Upheaval.

Subsidence.

Denudation.

any rainy day, trickling down a hill, or the springs which bubble up at the foot of cliffs on the sea shore, you will see an example of denudation in a small way. The earth is washed off the surface here and there, and carried down and laid up in banks in some places, and the harder ground underneath is laid bare. Little beds of stones are collected in one place, and sticks and straws and such light things in another, and this is just what has been done on a large scale in mountain regions, all over the world for many centuries.

In the uppermost sketch on page 51 you will see how the granite has been lifted up with the layers of other earth along its sides, and afterwards even layers have been deposited above; in the second there has been a great crack in the land, and a great mass of rock has subsided, and the hollow has become filled up in time with clay, and mould, and rich soil, so that some one has built a house and made a garden on it; in the third the river has cut a gorge in rocks which were once continuous from cliff to cliff, wearing away the softer earths more easily than the harder. If the Earth was cut into in different places we should find the rocks arranged in a very similar way to that in the three sketches.

BOILING SPRINGS, ETC.

In several different countries there are very strange sights, but scarcely anything is more astonishing than the fountains of boiling water which shoot up out of the ground. There are a good many of them not far from us, in Iceland, and many hundreds in Wyoming in America, and they are called "geysers." Steam and boiling water, and sometimes mud, are thrown up by these natural fountains to a height of 200 feet —as high as the top of the spire of a church. The water must come from a great depth in the ground—perhaps many thousand feet down—where the heat is intense. This water springing up with clouds of snow-white steam, and falling all round in showers, has a most beautiful appearance. These geysers now and then throw out very little water, just bubbling up above the ground, and then travellers boil eggs and chickens and such things in them, and have a pic-nic near them. It is impossible to say how long they have lasted, but we know from history that some have been spouting out water for at least 2,000 years, and how much longer no one can tell. They may have something to do with volcanoes, because water may have found its way to the heated interior of the earth, and being converted into steam, expands and causes an eruption.

Now that we have some idea of the construction of the Earth, we must go on to the *life* of the wonderful plants and creatures which have peopled it.

THE VEGETABLE PART.

THE DAWN OF LIFE.

The first beams of the rising sun, and the first grey light of the morning, tell us of the coming day; but we cannot even think of the dawn of that far-off day in the Earth's history, when no voice of man or beast was heard, and no trees or grass covered it, without solemn wonder at the immense distance that day is from us. A thousand ages are in the sight of the Creator but as yesterday, and the period of man's existence is only a moment compared to that of the lowly creatures which built up this World for him. In the first seas and on the land nothing was heard but the rushing of waters and the roaring of the fires of volcanoes.

It is impossible to be quite certain whether the first living things were animals or plants; but I think it most likely that very simple plants grew first, and that very simple animals came after or with them. Among the first of these, or perhaps the very first, were some small animals called *Eozöon*, which means the "life-dawn animal," and with them grew some simple plants. On the banks of the St. Lawrence river in Canada there is a great bed of rock called the Laurentian rocks, made almost entirely of the tiny remains of the "life-dawn animal," which, when we look at them through a microscope, are found to possess nearly the same structure as some lowly organized shells living in the seas now. These rocks are found in many parts of the world besides— in Eastern America, Bavaria, Scotland, and Norway; and in some places their thickness has been estimated at thirty thousand feet, or nearly six miles, or one hundred times as thick as St. Paul's Cathedral is high! These little creatures you see were at work over a great part of the Earth's surface, and you may fancy how many thousands of thousands of years it took them to build up these rocks. The "life-dawn animal" is far older than the chalk-building foraminifera, and so far as we know it lived alone in its seas. There were none of the beautiful twisted *ammonite* shell-fish, nor the shark-like fishes of the chalk seas. The eozöon was the only kind of living creature, the " lord of creation " for the time; and though storms raged in the seas it inhabited, the water was so deep that it lived on undisturbed. When you are able to use a microscope you will be able to see the traces left by these tiny animals in what is now hard stone.[1]

Life began in a very small way: there were none of the great land animals we have now; but these seemingly insignificant builders were at work so long that they made the immense rocks I have told you of. But this is not all. About this time some very simple plants grew on the land, and were carried down by the rivers and formed deep beds. After a long time these became covered up with different earths and were turned into the substance called "black-lead," which you use in drawing pencils. But this

[1] Specimen in Table-case 15, Room V., North Gallery British Museum.

is not really lead; it is almost pure carbon—in fact, the oldest kind of coal—so old that it will not now burn like coal, and is entirely made up of fossil plants crushed out of shape, so that we cannot now trace their forms, as we can the plants of the coal. When then you next take up a drawing pencil it will be easy to remember that the black substance which marks the paper was once a living plant, now changed by heat and pressure into almost pure carbon. As the name eozöon has been given to the "life-dawn animal," I will give this black-lead the name of *Eodendron*, or the "dawn-plant."[1]

Two very simple forms of life then occupied the earth and sea at the earliest time when anything at all was living, and strangely enough we use the dead bodies of both of them. We build houses of the rocks the eozöon laid down at the bottom of the sea, and the beautiful art of drawing is carried on with the carbon from the first plant life of the world—the eodendron.

[1] The name *Eophyton* has also been suggested for the earliest vegetable forms.

I must take you away presently to the coal, and sandstone, and chalk, and show you how plants and animals gradually increased in number and size, and fishes began to inhabit the seas, and all living things were slowly going on to greater perfection; for as time went on there was a steady progress from creatures like the eozöon, which had scarcely any power of moving about, to the active, quarrelsome and greedy things like crabs and lobsters which came after them, and the gigantic ferns of the coal beds. The peaceful "life-dawn animals" drew their food from the vegetable substances dissolved in the waters, though they perhaps also lived on animals still smaller than themselves; but, by-and-by, creatures, which must have been monsters to them, swarmed in the seas and devoured their smaller companions wholesale; and in time the Earth became very much the same as it is now, a place where the struggle for life is always going on. It is certain that animals have fed upon one another from the very beginning; but this is no doubt a wise law of the Creator to prevent them from increasing too fast, as they would do if all that were born lived, and none were destroyed.

We know much less about the vegetation —the plants and grasses—of the early ages of the world than of the animals; because plants rot away faster than bones and shells, and, besides, are less likely to be found in places where they would be preserved. A dead tree might be eaten up entirely by insects, as the white ants eat up fallen trees in a short time in tropical countries, and what is left of them crumbles away to fine powder and mixes with the soil. Immense trees are thus devoured now by millions of tiny insects no longer than your thumb nail, in India and Australia. No such thing as a whole and perfect fossil tree with every twig and leaf has been found; but then the coal beds are really great forests which have been buried for so long a time that they have quite altered in appearance. Still, among these coal beds we often find the bark, fruit, stems, and branches of trees very much like firs, and ferns, and huge club-mosses, which have the same shape they had when living, though they are quite black, and burn exactly like coal.

But there were plants long before the coal forests lived, and many fossil sea weeds are found in the old sandstones and limestones in Wales and other places.[1] The Old Red Sandstone, whose position you can see below the coal in the table of succession of formations, page 42, does not give us many fossil plants, though fishes and shells are common. This rock is found in Scotland, Herefordshire, Devonshire, and Ireland, as well as other places, and is often more than 2,000 feet thick. It was not all formed in salt water we know, because many of the fossil fishes and shells it contains are fresh water kinds. It must all have been made of the pieces of still older rocks worn away by rivers and settled like a sediment in immense lakes, some of which were fresh water. Then, after the Old Red Sandstone, came a time when the limestones below the coal were laid down at the bottom of a vast sea, and

[1] Divisions A and B of Case 1, Room I., North Gallery, contain some of the oldest known fossil plants.

here the remains of land plants are of course few. Then it seems there must have been a very long time when there were large continents all over the world raised above the seas, but not very much, and on these the forests grew which afterwards became coal fields. Until this time the plants had been mostly water weeds, reeds, rushes, and sea weeds, and it was not until England and Ireland became one continent, as they were once and covered with woods, that the great period of vegetation began.

The growth of plants was then most wonderful; but although coal is found in many different parts of the world, it was not all formed at one time, and though it is plentiful in England and Wales, Scotland, Ireland, France, Belgium, Russia, Hungary, Australia, New Zealand, China, and Borneo, it is older in some countries than in others. It is fortunate, however, that this useful material was made in Nature's workshop in so many different countries, or it would have to be carried from one to another. The coal forests were not the same trees as we have now—oaks, elms, ashes, limes, and so on. Most of them had rather hollow trunks and splendid waving tops like ferns and reeds, though there were some like our fir-trees.

If you lie down in the long grass before it is mown, and look through the stalks and fancy yourself an inch high only, you will have some idea how the coal forest would have looked if you had lived then. But there were no human beings on the Earth then, and I do not think there were any large animals, at least none have been found in the coal itself, except in Switzerland, where a few bones of the mammoth (an ancient elephant) and of the rhinoceros have been discovered in the much newer beds of coal, and also those of a large reptile like a crocodile in the coal beds of Ohio in America.

In such immense forests insects must certainly have been plentiful, and some of the fossil bodies of beetles, dragon-flies, and spiders, have been preserved, and a few tree lizards.[1] Of course the edges of the coal

[1] Fossil insects in Table-case No. 14, Room V.

II.

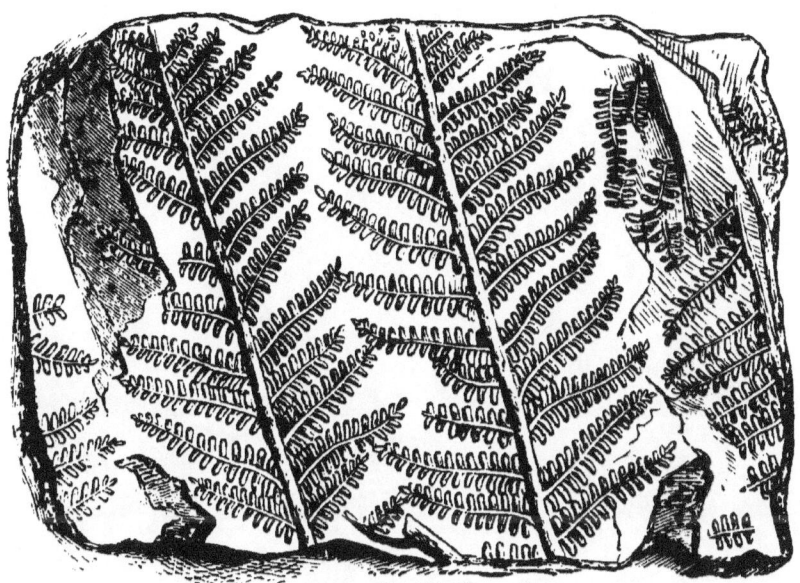

Fossil Tree Fern.

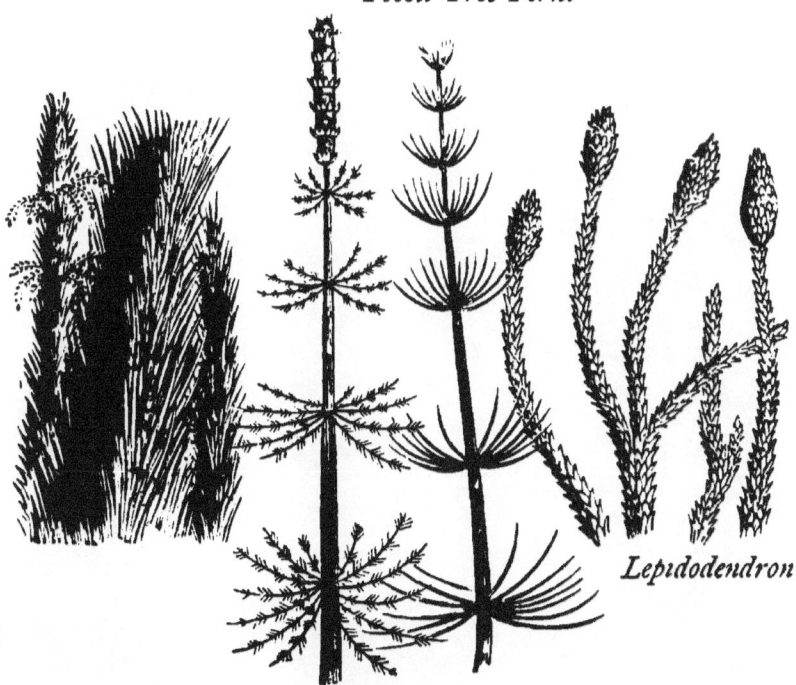

Calamites.   *Lepidodendron*
Different Kinds of Plants of the Coal Forests.

forests were washed here and there by the salt sea, and there must have been some fresh water rivers and ponds, for we find both fresh and salt water shells in these beds. It was almost dark in these forests, so thickly did the plants grow together. There were enormous club-mosses close together and as high as most houses, with their leaves interlaced making a complete network to shut out the sun. But the sun which shone on the forests was warm, and the air which went through them was soft, or they would not have grown so wonderfully. Indeed, there can be no doubt that the climate of northern regions was once much warmer than it is now. A thick bed of coal was discovered by the Arctic Expedition in 1875-6 actually within five hundred miles of the North Pole, where the ice on the sea is now thirty or forty feet thick![1] The forest which formed this coal could only have grown in a temperate climate, and there are no forests there now; it is so intensely cold they could not live. There must then have been a great change in the climate of the

[1] In 81° 44' N. latitude.

Arctic regions since that coal was living vegetation. The few plants and mosses which can live there now are of a very different and more hardy kind than those of the coal forests.

If you look at the engraving facing page 64, you will see a drawing of one of the tree ferns with its delicate fronds which grew so abundantly in the coal forests, and there are many other plants, some like the common "mare's tails," or *calamites*, growing in shallow ponds and ditches now—only the "mare's tails" or calamites of the coal forests were as high as poplars.[1] You can imagine what a splendid sight these forests of ferns, clubmosses, and "mare's tails," must have been, and what a multitude of beautiful insects and butterflies must have flitted about in them; but their frail bodies have almost all perished, so that we know very little of the animated creatures of the time.

Besides several sorts of coal both soft and hard there is a substance called "lignite,"

---

[1] Specimens of plants from the coal in Cases No. 2, 3, 4, in Room I.

which is scarcely wood and scarcely coal, of a brown colour. In fact, lignite is wood almost turned to coal, and it has helped us to learn that coal was once living wood; but it is not nearly so old as the coal. Then again there is the beautiful substance called "jet" used for making bracelets. This is a kind of fossil gum or pitch dropped from the trees while they were growing, and, though different in colour, it is much the same in kind as amber. Amber is often found with flies, spiders, and small leaves imbedded in it. When this fossil resin or gum was flowing out of the ancient pine-trees, and was quite sticky, flies settled upon it and became entangled in it, and as more of the gum flowed out they became quite covered. Then the gum dropped from the tree and hardened, and it is now found in lumps on the shores of the Baltic Sea, and in beds of sand and clay with fossil wood. It is of a beautiful bright yellow colour, and beads for necklaces and other ornaments are made out of it.

If we arrange the things we have been talking about in order, the oldest first, they would come thus : plumbago or black-lead—or, as I have called it, codendron, "the life-dawn plant"—first, then hard coal, then soft coal, then lignite and jet, then bog oak and peat. But I must tell you something about bog oak and peat. In many of the swamps and bogs of the World the trunks of dead trees are found, which have become quite black and almost like lignite, because they have been buried so long. Thus, in the bogs of Ireland oak trees are often found, and they were most likely living when the reindeer inhabited Ireland. This old bogwood is made into beads for necklaces and other ornaments. Peat is a partly decayed vegetable substance, with beautiful little plants growing on its surface, and is really coal in its infancy. It is found all over the world more or less in wet places, and consists of the roots and stems of mosses and reeds, some of which are like the gigantic plants of the coal period, but very small in comparison. I have no doubt that in time some of these peat bogs may be turned into coal if they sink down and become covered with other earths, but at present they are all on the surface and so soft that they are dangerous to walk upon because one may sink in and be smothered.

This, as far as we can trace it, is a sketch of the history of vegetable life on our Earth. We will go back to the coal for a moment and see what the animal life of that time was. The seas of the time of the coal forests were sometimes shallow, sometimes deep, and in the limestone rocks of the oceans which separated the great continents of that time there is a record of the inhabitants of the seas. The land plants were of more than 1,000 different kinds, and there were more than 200 kinds of fishes in the waters, and corals, shells, and small crab-like animals innumerable. The fishes were fellows with terrible teeth, and their bodies were covered with strong hard scales. One of these fish was thirty feet long, and there were others of considerable size. It is curious that the fishes of this time remind one of reptiles (lizards and crocodiles), just as the birds of a future time seem to have something of the reptile about them, as you will see by-and-by.

I dare say you have remarked while reading that all the plants and animals of the early ages of the world seem to be made on a simple plan, and as the Earth grows older they become more perfect, and this is just what I want you to take notice of all through. The plants of the coal period, you have seen, were nothing like so perfect in construction, beautiful as they were, as the forest trees of the present time, neither were the animals so perfect as those living now. There has been *progress*, step by step, throughout the vegetable and animal creation ; and, though many of the lower forms of the early ages exist now, there are others far superior to them which did not exist then : but all this will come in "The Animal Part."

About the middle of the Earth's age came the wonderful period of vegetation which gave us our coal, and after that there was a great and busy time, when huge reptiles and reptile-like birds, and then true birds, made their appearance. But that belongs to the next part of the "puzzle of life."

If we look with astonishment at the coal

forests, we may also well think of them with thankfulness. Here is the sunshine of past ages stored up for our use, and we bring it out again to warm ourselves, cook our food, make all our iron things, and drive our steam-engines! Can any romance be finer than this, that we are carried across to America and India and Australia in steam-boats driven by the "fossil sunlight" of ages and ages past, and whirled along at sixty miles an hour over iron rails by the same stored-up strength?

If you doubt this, think of living trees. Do they not live by the air and sunlight? Will they grow without these? They spread their branches and leaves to gather the warmth and light from the air, and when they are cut down and dried, and you put a match to the wood, all the old warmth and light come out again; and we know that the coal is only fossil wood. Our Creator wastes nothing. Even when there were no people living to rejoice in the sun, He thought of those people who *should* come in time, and not one of the fiery rays of the fierce sun was lost. These mighty forests were sent to gather it, and when they had died down they sank below the surface and were covered from the air, that none of their light or heat should escape.

In such forests it is strange that there were no birds, especially as there were swarms of insects, and no doubt abundance of worms. But no bone of bird or any trace of feathered songster of these lovely groves has yet been found. Little lizards chased flies and beetles up and down the stems of the club-mosses and ferns, and larger reptiles lurked in the long damp grass under the shade. The pools and ponds were filled with curious fishes, and reefs of beautiful white coral fringed all the shores of the seas.

But the Earth was not fit for the habitation of man. The fruits of the trees were not such as he could have eaten, and their wood was not hard enough to build houses of. Still it was being got ready for him, and not a leaf waved uselessly in the bright, warm air, and not a tree fell to the ground, but it was to be turned into coal, and to come forth again one day a hard black lump, without any of its former beauty, but to give back the light and heat it had gathered from the sun ages and ages ago.

Many periods in the Earth's history have passed since the coal period, and in every one of these the trees have been increasing in perfection, though there have never since been such great numbers of a few kinds growing. When we come to the more lately formed beds of earth we begin to find the cypress, willow, ash, oak, elm, and other forest trees which are living now. The trunks of these trees, blackened by age, lie buried in peat bogs and swamps all over Europe. The mighty Mississippi river brings down immense quantities of dead trees, and as these sink to the bottom near its mouth they are forming future coal beds. Along the coast of Norfolk and Suffolk, too, and stretching far away under the German Ocean, is an old English forest. In some places the trunks of the buried trees may be seen standing upright just where they grew. The nets of the fishermen are continually bringing up pieces of wood, roots, and seeds; and when the sea washes away the soft cliffs here the bones, teeth, and tusks of the elephant, rhinoceros, hippopotamus, and other large animals which inhabited this forest, may be seen in great numbers.

Down below the waves of ocean have these woods sunk with all their once living creatures, and though you may suppose that it must have been very long ago that they grew, they are of the same kind as those which now make the hills and valleys of England beautiful.

Sometimes a forest must sink very fast, for travellers have told us how they have sailed on rivers and lakes over the tops of sunken trees, and, looking down into the clear water, have seen the branches waving below—tall trees standing upright at the bottom, and the boats sailing over their tops!

We must now pass on to the living creatures which peopled the Earth, and their story can be told with more certainty than that of the perishable plants which clothed the surface of the ground, and, while they rendered it beautiful, also served as food and shelter for innumerable animals, and have become so useful to us as coal, lignite, black-lead, and other productions of ancient forests.

THE ANIMAL PART.

WE must now go back and collect the smaller pieces of "the puzzle" which make up the animal part. The great periods of vegetation ended in our country with the coal forests, and there has been no such wonderful growth of plants since the time when the New Red Sandstone, lying above the coal, was formed; though no doubt trees and plants have since flourished, as they do now on the Earth, but not in such quantities as during the coal period.

We remember that the eozöon, "the life-dawn animal," is the oldest animal we know of, and that it lived so long ago as when the Laurentian rocks were laid down at the bottom of the seas of that time; then in later rocks we find the burrows of sea worms in the stone, and later still simple shells with two valves like the common mussel, and other animals of a simple kind, like the corals, sponges, and star-fishes which exist now. There must have been millions of these creatures in the older limestone seas, for the rocks are almost entirely composed of their fossil shells and bodies. By-and-by a rather superior animal inhabited the seas of Wales, called a trilobite, of which you will see a picture on the opposite page. This curious animal was of the same family as the shrimps and prawns, but much larger, and he must have been a giant among the others. None of these animals had any bones, you must understand; but they had a hard shelly covering to support their soft bodies inside, and no doubt the trilobites were able to swim about very fast.[1]

What I want you to take notice of now is the *progress* that has been going on from the almost motionless eozöon to the shell-fish and star-fish, which could crawl along the bottom of the sea and over the rocks, to this active, quick-moving trilobite, with his great paddles. Then the next step is a very great

[1] Numerous specimens in Case No. 7, Room V.

III.

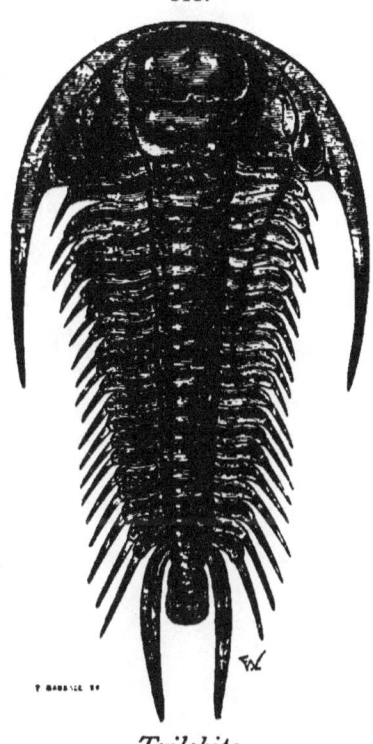

Trilobite.

one, when we come to animals with bones. The first of these are fishes. All the other bones are joined to the backbone, therefore all animals with bones are called *vertebrata*, which is a Latin word meaning having a backbone with joints. Now animals with bones are plainly superior to those with only shells, and when we find fishes among the rocks of Wales and Devonshire we know that we are beginning to pick up some important pieces of the "puzzle of life." These fishes were most of them related to the sturgeon, and their bones and teeth are found in great quantities in the Old Red Sandstone rocks, just below the coal.[1]

It is not until we get above the coal into the oolite or egg-stone rocks that still larger and altogether superior animals, both of sea and land, began to increase, and this is called

THE AGE OF REPTILES.

This has been called the reptile age because there were such numbers of animals

[1] Specimens of fossil fishes from various rocks in Wall-case No. 1, Room II.

like crocodiles, lizards, and tortoises (which are all reptiles), and some of them were of immense size. For instance, there was a huge creature something like a frog, but as large as a Shetland pony, called the *Labyrinthodon*, with a great many curious teeth, and this animal has left footprints in the New Red Sandstone which have been dried and buried, we can't tell how long, and there are the cracks made by the sun drying the place he walked over when that was soft earth. There is a drawing of some of these footsteps in the picture on the next page, and there are also the footprints of a large bird, and you can see where he walked over the soft earth and made a long line of footmarks; and if you look at the footprints of birds on the snow or mud now you will notice marks just like these. Then there is another picture of a single footprint of a large bird, and all those round dots are where rain drops fell and left their marks in the soft earth.

I dare say you will wonder how it is that these footprints have not disappeared. Well, when the animals and birds that made them

IV.

Footprints of Labyrinthodon.

Footprints of Birds, (2) with marks of Rain-drops.

had gone the marks became filled with dry sand, no doubt blown in by the wind, and then the mud dried hard, and at last it became covered with other earths and sank slowly down, just as the coal forests had done before, and remained there until we dug it up with these tracks of the birds and animals that lived then. Some of these birds must have been larger than any living now, because their footmarks are so long. None of their bones have been found yet, I believe, but plenty of the teeth and some bones of the labyrinthodon have. The real footmarks, of course, are very large, though they are small in the picture.[1]

In the great beds of Lias there are many other strange animals, and among them are two great fish-lizards called the *Ichthyosaurus* and *Plesiosaurus*. Both of these lived in the water and perhaps came on land sometimes, and it is certain that they must have been very ferocious creatures, from their great size and sharp teeth. The plesiosaurus would be able

[1] See examples in the large Wall-cases in Rooms I., II., and III., North Gallery.

to raise his long neck above the water and snap at some of those curious birds rather like bats which lived at the time, and of which I shall have something to say presently. Some of these fish-lizards were as large as whales, and their bodies have been so beautifully preserved in the limestone rocks that we can actually sometimes find in their stomachs the food they lived on.

Now we have got to a higher order of creation still, these fish-lizards, and they remind one of the next step in progress—birds. You know that all birds lay eggs, so do almost all reptiles, such as crocodiles, lizards, and most snakes, so that they are alike in this. Then the plesiosaurus with his long neck reminds us of such birds as the heron and the swan, but he is altogether more like a reptile than either a fish or a bird. There were also huge land reptiles, which lived in the forests of the time, and must have been a terror to the smaller animals. From the bones of one of these which have been found in the oolite clays near Weymouth in Dorsetshire (the *Cetiosaurus*), we see that it must have

V.

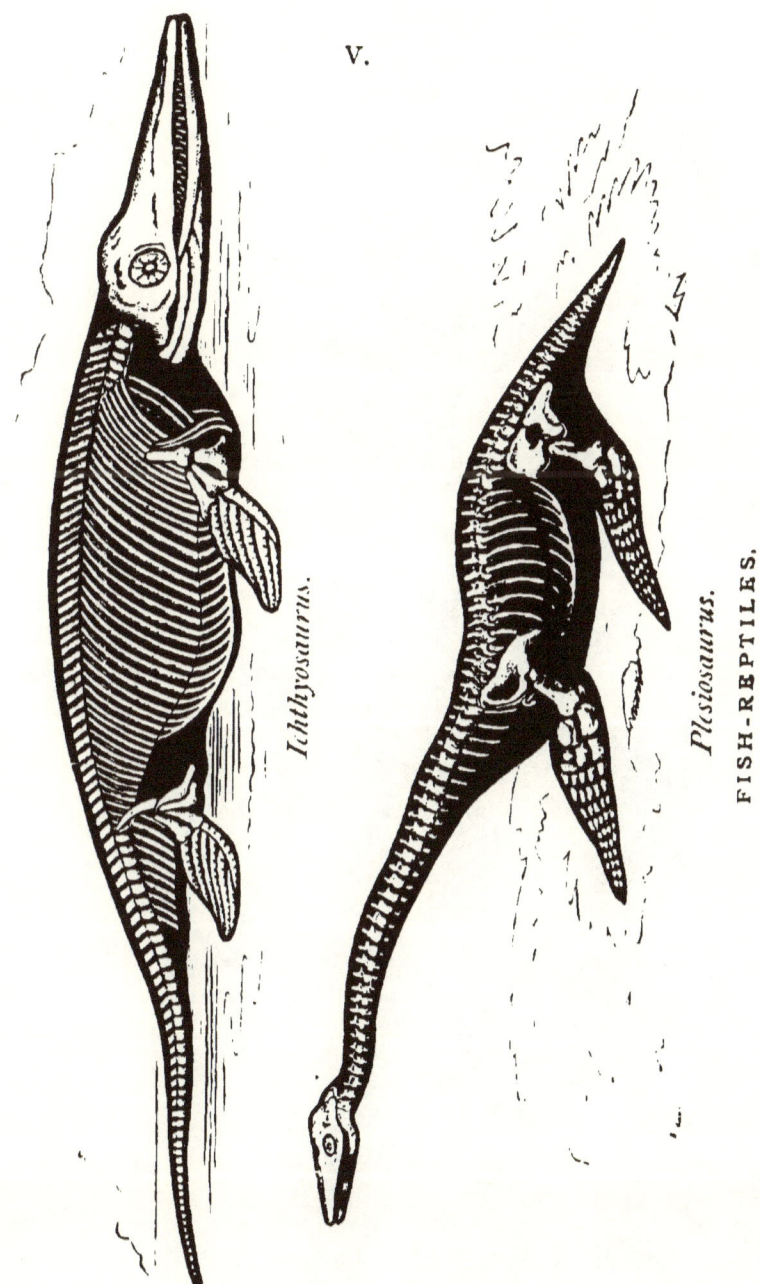

Ichthyosaurus. Plesiosaurus.
FISH-REPTILES.

been nearly as large as an elephant, and there are others called the *Megalosaurus*, *Dinosaurus*, &c. All these names end with *saurus*, a name taken from the Greek word meaning lizard ; and you will see now why the oolite, or "Jurassic"[1] age, as it is sometimes called, is well named the "reptile age," for these creatures swarmed on the land and in the sea. Specimens of these you can see for yourselves in the cases on the walls of the third room in the North Gallery of the British Museum, where all the fossils are collected.

But still more extraordinary animals than any of these lived at the time, and we can scarcely tell whether they were birds or reptiles, as they were something like both, but I suppose we must call them flying reptiles, and they are the nearest approach to birds that had yet existed. These creatures are called *Pterodactyles*, from two Greek words which mean "wing-fingered." Suppose the little fingers of both your hands were a yard longer than the others, and suppose a thick

[1] So called because the mountain chain of the Jura Alps was raised during this period.

leathery skin was stretched from the tips of your long little fingers to each of your feet, you would have wings something like a pterodactyl and also something like the wings of a bat. But the pterodactyl had a long neck and a long beak-like mouth, full of long sharp pointed teeth. It could not walk much I think, but it could hang itself up by its hind limbs to a tree or rock, head downwards like a bat, and must have been able to fly very strongly, with its huge leathery wings, but it had no feathers. There were swarms of these curious half lizard half bird-like animals on the land, and they were of all sizes, some no bigger than a crow, and some as large as the albatross, measuring twelve feet across their outstretched wings. Their skeletons are some of the commonest fossils in the oolite rocks, all through the great reptile age.[1]

Now you see we have come to a reptile that can fly, but, excepting for its wings and some of its bones, more like a crocodile than

[1] Several specimens in Room III., and in Table-case No. 16, Room IV.

a bird. A little further on we find another curious animal in the oolite rocks, which is much more like a true bird than the pterodactyl, because it had feathered wings. It is called the *Archæopteryx*, which means " ancient wing," and I have given a picture of it on the same page as the pterodactyl, so that you may compare them together. The blade-bone and " merry-thought " of this creature were exactly like those of a bird, and so were the feet and legs, which would enable it to walk easily, or perch on the branch of a tree, but the tail was long and many-jointed like that of a lizard, with a fan of feathers growing on each side of it, and short feathered wings. Then it most likely had teeth like a lizard, and there were short claws at the bend of the wings. This bird-reptile was about the size of a crow, and was the first we know of with feathers, and the limestone rock has preserved it most beautifully through all the long ages which have passed since it flitted over the land of the oolite period.[1] Later still than these, there

[1] Wall-case No. 11 in Room III., several specimens, imperfect.

lived in America, about the time the chalk was formed in England, two strange birds called *Hesperornis* and *Ichthyornis*, both of which had teeth in the jaws. The former was an immense fellow like the penguin, with short wings, and the latter was about the size of a pigeon with large feathered wings.

They are finding more of these curious creatures every now and then in America. Some are without teeth, and have a horny bill like that of a real bird, and in other ways more nearly resemble living birds; still they have not lost the appearance of reptiles in their principal bones.

I have been particular in describing some of these fish-lizards and bird-reptiles; because they, or their near relations, were the principal inhabitants of land and sea from the end of the coal period to the end of the chalk, though there were of course swarms of fishes and shell-fish; but I ought to tell you that even so early as this there was at least one animal known which suckled its young ones, and this was a small insect-eating creature not larger than a rat, of the same family

VI.

Pterodactyl (Wing-finger).

Archæopteryx (Ancient-wing).

(called *Marsupial*) as the kangaroo of Australia, which carries its young ones in a pocket or pouch in its skin.

All this time we have been hunting for parts of "the puzzle" in those ancient oolite rocks between the coal and the chalk, and those we have found are very important. We have seen the slow progress from simple sea shells to simple fishes, and then onwards to fish-lizards and bird-reptiles with one little marsupial animal, of a far higher kind, in between, as if to tell us beforehand what more complete and perfect animals we might expect by-and-by. After the fishes we have found fish-lizards, then bird-reptiles with wings, but no feathers, and later still a bird-reptile with wing and tail feathers. How different the life of the Earth was at the end of the "reptile age" of the oolite rocks, to the far back Laurentian time when one little creature, our old friend eozöon, alone held possession of the seas!

THE CRETACEOUS PERIOD.

Now let us look into the rocks next above, and see what is to be found there.

We have arrived in the Cretaceous period, or time when the chalk was formed.[1] You remember I told you you might call this "foraminifera earth" because so much of it was made up of the shells of these tiny animals, thousands of which could be put into a thimble. Whenever you make a mark with a piece of drawing chalk you rub off a number of them, and you will see what pretty little creatures they were if you look at the drawings of some of them on the next page as they are seen under the microscope, magnified thousands of times their natural size; but there are others of different shapes. On the same page too there is a handsome shell, called an ammonite, and of its real size, common in chalk rocks. The seas of the time must have been very deep as I have explained before, and the chalk contains numbers of bones of fishes everywhere, and many of the remains of the reptile-like creatures of the time before. Corals, sea-urchins, crabs, &c., abounded, and as you can scarcely ever see chalk without immense flint stones in it, you may suppose

[1] From the Latin word "creta," meaning chalk.

VII.

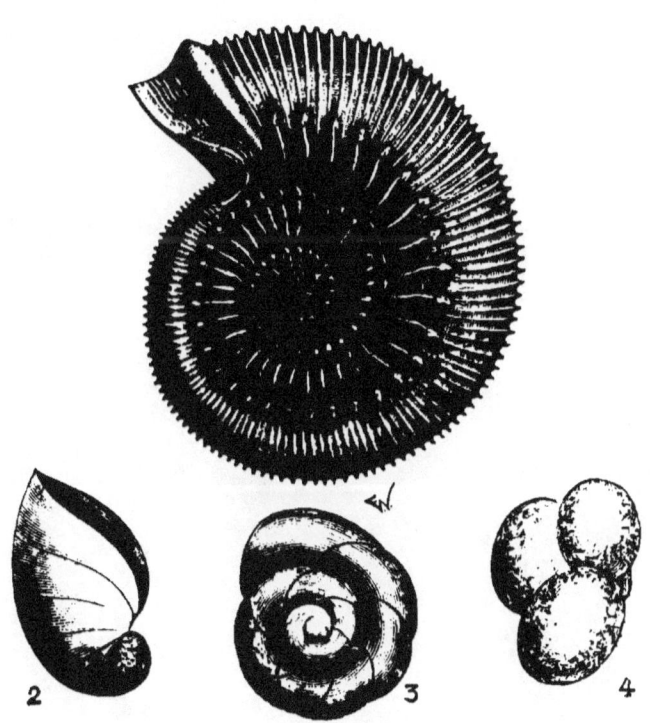

FOSSILS OF THE CHALK.

1 *Ammonite.* 2 3 4 *Foraminifera (Chalk-builders).*

what millions of sponges lived on the rocks, for these flints are partly made up of their fossil bodies.[1] Another Cretaceous period is beginning now at the bottom of the Atlantic and Pacific Oceans, where it is deep enough to cover the Alps, for these little foraminifera are living on the surface in countless millions, and day by day their fossil shells are settling down to the bottom and forming a soft grey mud, full of the carbonate of lime like chalk. The climate of the Cretaceous age was mild and pleasant, as we know from the kind of animals in the seas. Slowly the water began to get shallower and shallower by the upheaval of the bed, and at last the bottom of this mighty chalk ocean came up to the light and sun, to be covered in some places with the drift and worn particles of older rocks swept over it by rivers, and to receive new plants and new animals, and in some places to remain almost bare, as it is on the downs of Brighton.

Now we take one more step upwards into

[1] Ammonites in the Table-cases in Rooms V. and VI. For enlarged models of foraminifera, see Case No. 15 in Room V.

almost a new world—the world on which mighty animals lived, and which man came to share with them.

THE TERTIARY PERIOD.

The reign of the reptiles is now passed. The ichthyosaurus and pterodactyl no longer inhabited the seas and continents. Great changes had taken place in the shape of the land. A river larger than the Rhine swept majestically through England from the borders of Wales right out into the German Ocean, and its banks were covered with forests and marshes, where the new animals which had come to take possession of the earth lived and moved and had their being. The mountains of the Pyrenees were raised above the sea, and parts of Surrey and Sussex appeared too. It was most likely in the early part of the Tertiary period that the stone was formed of which almost all Paris is built. Fancy a great city built of the shells of dead animals! One can scarcely believe it: but the microscope lets us into this secret of Nature. If we take a piece of this stone and examine it in a powerful microscope we see that it is made almost entirely of tiny shells, so small that myriads of them could be packed in a nut-shell. How long must they have been working to make all the stone beds of which Paris is built? We cannot measure the time, we can only know it must have been enormous!

All kinds of animals both of sea and land increased in numbers and perfection. The ammonites were dead, but their even more beautiful relation, the nautilus, was living as it is now. The trilobite was gone, but his next relation, the lobster and crab, appeared. Fishes abounded. Whales which suckle their young ones appeared, and the numbers of vertebrata, or animals with backbones, were more numerous than they had ever been before. Just as animals with bones are more perfect than those with only skins or shells, so animals which suckle their young ones are more perfect than those which only lay eggs. Thus the whale is a more perfect animal than the shark, though both inhabit the water; and elephants and even rats and mice more perfect still; and because there were so many of these "sucklers," or mammalia as they are called, in the Tertiary period, we know that all living creatures were becoming more perfect. It will interest you too to learn that monkeys began to appear now, and that they were common in France, while at the present time the only part of Europe where they are to be found is on the rock of Gibraltar.

But I want particularly to tell you of the giant animals—the Mammoth, Mastodon, Megatherium, Dinotherium, and others, and first let us see what the mammoth was like.

In former times, when people accidentally found the bones of these animals, they actually thought they had belonged to giant *men*, and we can scarcely wonder at that : but we know better. If only one small bone is shown to Professor Owen or Professor Huxley, he can tell at once whether it belonged to a man or an animal, a fish or a bird, and very often the particular animal too. Well, the bones of the mammoth were found in the north of Russia on the banks of the river Lena in 1800 : but the Russians knew of them before that, and the name they gave the animal means "earth," because they supposed it burrowed in the earth like a mole. This one is now in the Museum at St. Petersburg, and its brownish coat and long black hairs, and even the hoofs and some of the flesh, can be distinctly seen. The drawing in the frontispiece is taken from it. It was strange that any people could have supposed that this huge creature, larger than an elephant and with great curved tusks ten feet long and weighing 160 lbs., could have got underground of its own accord : but that was the only way in which they could account for finding it buried in the earth on the banks of the rivers. Look at the picture in the frontispiece ; what a splendid animal he was, this old elephant ; larger and stronger than any living elephants! Immense quantities of their bones are found in Siberia, and the tusks and teeth are brought in ship-loads to England, where they are sold for their ivory. Their skeletons have been found in most countries of Europe, in many parts of Asia, and in North America, and these animals must have been common at one time near London, for their bones have been dug up in the brick earth at Ilford in Essex and other places near the Thames. There is a skull with tusks set up with iron supports in the British Museum.[1]

There was besides another animal very much like this called the Mastodon; but it had tusks in the lower jaws as well as the upper, four in all, and the lower tusks dropped out when the animal grew old. The whole skeleton of one of these is also put up in the Museum, which you ought to go and see.[2] Mastodons' bones have been discovered in England and other parts of Europe, and in North and South America and India, so that they were spread pretty well all over the world. They had very curious pointed teeth rather like a lot of fir cones piled together, not flat grinders like those of the mammoth and all living elephants, and perhaps they fed upon fruits and nuts, and boughs, as I do not think they could have managed well to chew grass and leaves with such pointed teeth. The teeth in their old dead jaws are still beauti-

[1] Room VI., North Gallery.   [2] In the same room.

fully white and look like china. Both the mammoth and the mastodon had long trunks of course, and they must have been grand looking creatures marching about in the English forests. We should be very much startled if we were to meet one of them now in an English wood : but there is no chance of that, they have all passed away, and the only relations they have living are the elephants of Africa and Asia.

During this Tertiary period, or at least the early part of it, besides the mammoth and mastodon, the hippopotamus and rhinoceros were plentiful about the Thames. Those same Ilford marshes in Essex have been a complete storehouse of the remains of these animals. The bones of a hundred different mammoths and eighty rhinoceroses have been dug up lately from the damp, black soil, as well as many belonging to the hippopotamus, and we can have no doubt that all the swamps along the north side of the river were inhabited by large herds of these huge beasts, or so many of their skeletons could not have been collected in one place. It is very likely they were overtaken in a flood of the river and drowned, and their bodies sank down in the mud of the river bank : but anyhow, there they are to tell us that they lived and died almost within sight of the Tower of London, if it had been built then, as of course it was not.

Long long ago too, before there was a single brick where London stands, and when the few human beings who were living were obliged to hide themselves in caves, great lions might have been heard roaring at night in the forests of the Thames Valley. The bones of this lion have been found in many different parts of England, and a terrible fellow he must have been, for some of his canine teeth (the long sharp teeth in cats and dogs) were more than six inches long. Indeed they were like small swords, and this is why he has been called the "sabre-toothed" lion. There were also bears, like the great grisly bear of America, and leopards, hyenas, and wolves, and besides two kinds of ox far larger than those we have now. But one of the handsomest animals was the great Irish stag.

VIII.

Gigantic Irish Stag (Cervus Megacerus).

When standing upright the top of his horns would be as high as two tall men. He was indeed a fine fellow with his immense spreading antlers. The deer in our parks would look dwarfs beside him. He inhabited both England and Ireland: but, being found more often in Ireland, he has got the name of the *Irish* stag. As many as thirty of the skeletons of these stags have been found together under a bog in Ireland, and in some of the bones the marrow is still preserved, and they burn well. Fences have been made of these bones in Ireland, and when the people of a small village in the county of Antrim heard of the battle of Waterloo they made a great bonfire of the bones and horns of the Irish stag to rejoice over the victory. I dare say these stags were hunted by wolves, and perhaps driven on to the ice of ancient lakes, where they broke through and got drowned, for so many of their skeletons are found together. I could not pass this magnificent stag by without giving you a picture of him.[1] He

---

[1] Complete specimens of male and female in the middle of Room V.

was a much nobler looking animal than the reindeer, which lived along with him at the time in England, and from his appearance I should say he was a swift runner and great fighter. Some antlers have been found locked together, just as these stags died in mortal combat, and I never see Sir Edwin Landseer's beautiful picture of two red-deer stags fighting without thinking what a grand sight it would have been to see two of these great Irish stags rushing at each other with their powerful horns.

Not one of those animals is living now, and none of them is mentioned in any history or tradition whatever, and though there is no doubt that men living in Europe saw the mammoth alive (as you will find in the next chapter), they knew of no kind of writing in which to tell us of them; these fossils are the only records left, but they speak plainly enough of the time when England and the whole of Europe were inhabited by these races of huge animals.

Now I must carry you away to South America, where there are more wonders. If I were to tell you of all the singular monsters

IX.

people have found in the beds of the rivers there it would make a book of itself. You know what large rivers there are in that country, and how they run for thousands of miles through almost flat plains called "Pampas." Well, these rivers have often changed their beds by cutting new channels in the soft soil. The old dry beds of the rivers are the burying-places of some most curious animals, but I have not room to tell you about more than one of them at present. He is called the *Megatherium*, which means "great beast." His size and strength were enormous. The largest hippopotamus looks small by his side. His leg bones are bigger than your body. He was more like the sloth than any other living animal, but he could not climb. He stood on those huge, broad hind feet, with his strong tail as a sort of third leg, and tore down the branches of the trees to feed on, or even rooted them up to get at the leaves. Standing by his skeleton in the British Museum[1] one feels quite a shrimp, and he looks strong enough to walk

[1] Room VI.

away comfortably with an elephant on his back.

Another immense animal inhabited South America at the time, which geologists have called *Dinotherium*, or "dreadful beast."[1] He was a relation of the mastodon, but his tusks were very curious. Instead of being in the upper jaw and turned upwards they stuck out from the lower jaw and curved downwards, giving him a very odd appearance. He most probably had a trunk like the mammoth or mastodon, but perhaps not so long. All these of course were vegetable feeders.

The Tertiary period is so remarkable for the numbers of animals more or less related to elephants and spread all over the world, that we might almost call it the "elephant age," as the oolite has been named the "reptile age." These elephantine animals abounded in Europe, Asia, and North and South America, and though none of this kind have yet been found in Australia and Africa, I cannot help thinking they will be discovered in Africa at all events, for there is no doubt that Africa and Europe were once joined.

[1] Head and tusks in Wall-case No. 2, Room VI.

Australia you know possesses that animal so unlike all others that when we first see it we are quite astonished—the kangaroo. The bones of a huge fossil kangaroo have been found in Australia which must have stood fourteen or fifteen feet high I should think when on its hind legs, or more than twice as large as any living now.[1] Then there were giant birds in New Zealand (something like the ostrich) called *dinornis* or "dreadful bird." These fellows had no wings, and they must have been very much taller than the ostrich or emu. To look at their leg bones you would think they were the bones of oxen instead of birds, they are so immensely thick and strong. I do not think any of these are living now, because they have been sought for carefully, and none of the natives even can say that they have seen one. But their skeletons are common in the surface earth, and their bones, cracked to get the marrow out of them, are often dug out of the heaps of refuse collected about ancient cooking places. So that they were used for food, and

[1] Skull in Wall-case No. 1, Room VI.

perhaps they have not been extinct—that is to say, died out—more than a few hundred years; and this is more likely because feathers are sometimes attached to the remains, and undecayed sinews on the feet. A human skeleton has been found in a grave in New Zealand, too, with the egg of one between its arms, and little piles of pebbles are often seen among their bones, where the stomach would be, which the bird swallowed to digest its food, just as many birds do now. The natives called it the Moa, and they have some traditions about it, and, all things considered, it is probably one of the most recent fossil animals, and that is the reason why I have left it to the last.[1]

Now I dare say you will wish to know when the animals living now took the place of those I have described, and which have all passed away. This cannot be told with certainty, but you will see in the "Human Part" that Men were living when the mammoth, mastodon, and some other extinct animals, inhabited the Earth, and that the reindeer, ox, bear, wolf, hyena, &c., have survived to the present day.

[1] Several specimens in Wall-case No. 11, Room III.

Throughout these immense periods of time there are gaps which we cannot yet fill up. No one can yet say, for instance, when the last of the mammoths disappeared, and the first of their near relations, the Indian and African elephants, took their place. These are the missing parts of "the puzzle of life" which you may perhaps one of these days find when you come to study the subject, and when you have learned all that is known at present. But you may be sure of this, that throughout all time there has been *progress*, the lower forms of animal life have been followed by more perfect forms as the Earth grew older. It is true the lower forms of life have not all died out. These imperfect animals have run through all the ages—the chalk builder of the Cretaceous age lives in the ocean now—and there are many other simple animals which lived in Old Red Sandstone times, and are not extinct yet, but wherever a superior kind of animal has passed away another more perfect has taken its place. This will be seen at once if we compare the "Reptile Age" with the Tertiary. The great ichthyosaurus, plesiosaurus, and pterodactyl are gone, but now we have the more perfect crocodiles and birds. The mammoth is gone, but we have the elephant. There are no giant mosses or towering tree ferns, but our forest trees are more perfect and more varied. The plants which formed the coal forests and once clothed the Earth with beauty have dwindled away to the lowly forms which we must stoop to examine in swamps, and these humble plants are all the surviving relatives of their once noble family. The lordly oaks, and elms, stronger, and even more lovely in the sweet drapery of their foliage, and much better fitted for our use, have succeeded all those soft-stemmed plants which grew so fast and were the best possible kind for forming coal.

When you are able to study what is called comparative anatomy you will see how wonderful the *plan* of creation is, and how beautifully it has been worked out by its great Designer. You will see in the bones of the reptiles of the oolite rocks a prophecy as it were of the birds and animals which were to come. What could be more prophetic of animals with the power of perfect flight than the leather-winged pterodactyl, half lizard and half bird? In some of these animals you will see bones only half formed, and useless to that creature, which were brought to perfection in later times, and became the most important part of the body.

It is very difficult for me to make all this plain to you, but if you are really interested in it you will go to a museum where the fossils are collected, and then I am very much mistaken if you do not find a new and strange world opened to you.

THE HUMAN PART.

THE history of the human race is of course even more interesting than that of the plants and animals which lived so long before man and prepared the way for him, because man is the "crown of creation."

When first placed on this Earth he must have been but little superior to the animals in his outward life, though he had very different powers within him. He could gather the fruits of the Earth like them, and perhaps used some of the smaller creatures as food, but he could do little more. He scarcely knew that he possessed the faculties which would in time make him lord of the Earth and the creatures inhabiting it. By slow and painful experience he was to gather those stores of knowledge that were to enable him to overcome difficulties, to provide him with shelter from the weather and protection from dangerous animals, give increasing comfort and power, and set him so far above all other created things. He found plants and animals for his use, and the dwellings in caves and holes ready made by Nature. He could neither build houses nor make weapons. The first weapon he ever used probably was a stone, which he could throw at small animals. Then he would find out that long, sharp-pointed sticks could be thrown like spears, and he also found that a long pliant piece of wood when bent would fly back, and in this he would see a means of throwing smaller pointed sticks like arrows, and I dare say the discovery of the way of making a bow with a string of twisted animal skin was a great invention, and it certainly would be a very valuable one. Many generations must have passed away before he got even as far as this. It is very easy for us, who see bows and arrows from our childhood, to understand their use at once : but the first human inhabitants of the world had to find them out for themselves. They began with *no* knowledge at all. The beasts of the field and the fruits of the Earth were given them, but they could MAKE nothing.

They had not even the natural covering of hair, or wool, or feathers, which animals and birds have, and they must first have clothed themselves with skins of these. The wants of their daily life were so great that they had no time to think of anything else, but when it became easier to satisfy these bodily wants their minds turned to other things. They must have seen that when the seeds and fruits of plants fall upon the ground they grow and produce the same kind of plant, but they did not at first think of gathering a great number of these seeds and sowing them in one place and making a garden. They could wander about and gather all they needed as they became ripe, for there were few people then. Their life was like that of the lilies of the field, they " toiled not neither did they spin," as Christ says of the flowers, but when they began to increase in number something more was wanted. People began to feel something within them which we call "intellect," and this must be satisfied. It was not enough to live as if they were no nobler than the animals. Something stirred in their minds which told them they must not stand still.

The Creator has made both us and the wood and stone and metals, and has given to us the power to make other things out of them. Thus we are nearer to Him in power than any of the animals who cannot change the rough materials into other forms. We admire the simple and really beautiful nest of the bird, but we feel that our power is greater when we consider our splendid buildings and steam-engines, our ships, and our many conquests over difficulties. But if we did not use these greater powers of mind and hand well, we should find them grow weaker and weaker until we might almost lose them.

You may easily suppose that there was a time when men could not write, and there were no books of any kind, nor any other means of exchanging thoughts except through spoken language. The earliest histories about the human race always speak of men who lived before those histories were written. We have nothing about the earliest men written by *themselves*. It is always someone else who writes of them, referring to their deeds, and to events which happened long before.

The art of writing has grown up gradually and very slowly, for when the inhabitants of the Earth became numerous they felt the need of some way of expressing themselves to those at a distance from them, and for making a record of things that happened and might be forgotten. Some of the earliest means of writing were by pictures, like the picture writings of Mexico[1] found by the Spanish conquerors, and something of the same kind is even now used by the Chinese and Japanese. Their writing is made up partly of pictures and partly of queer signs which stand for the names of things, as you know if you have ever seen one of their books. One of the oldest forms of writing known is the hieroglyphic, which is said to have been first used by the Egyptians about 2,100 years before Christ, and another is the arrow-shaped writing of the Assyrians. These were cut on stone and metal tablets, and most of them are the histories of their kings. But there are some writings on stone in India which are thought to be older still. The Egyptians

---

[1] A fine Mexican MS. on diapered cloth, with figures and mystical signs, has lately been added to the MS. department of the British Museum.

made great progress in writing afterwards when *papyrus* was invented.[1] This is a kind of paper made from a reed which grows abundantly in the river Nile, and many of these papyrus writings are preserved in the British Museum, as well as the writings on stone of the Egyptians and Assyrians, and learned men have spelled out a great deal of the history of these nations from them, though the language is quite different from any spoken or written now.

Picture writing was most likely one of the earliest inventions in this way: but it was so troublesome that signs were used to express the same things as the picture. For instance, suppose a history of a king was to be written. The word "king" would be shown by something he always wore, such as his crown, and this sign would become more simple until at last it might not be anything like a crown; but it would be remembered that the sign stood for a king all the same. The first letter of the Hebrew alphabet, *aleph*, means an ox, and the letter is something like

[1] Some fine examples of papyrus writings on the Northwest Staircase, Upper Floor.

the shape of the head of that animal with its horns; and another letter, called *shin*, which in Hebrew means a tooth, is actually very like a tooth with three points. In many languages these signs have become so altered that they do not now resemble the things they at first stood for ; but the first steps in the invention of written language were certainly made by signs representing the thing of which the person wished to give an idea. But you will learn all about these ancient writings from other books.

The men whose lives I am going to describe lived long before any of these writings were invented. They *spoke* a language of course, though there is nothing left to show that they knew of any kind of writing, and they are called Pre-historic men because they lived before there were any histories either written by themselves or about them. But they could draw a little, as we know from the pictures of animals, birds, and fishes scratched upon pieces of slate, and bone, and stone found in their graves. Perhaps these pictures were memorials of their great or wise men, or showed that they were clever hunters, or fishermen.

They knew the use of fire. Half burnt bones and wood and ashes are plentiful in the caves where they lived. They had none of the means we possess for kindling fire, and there are only two ways by which they could have got it. They might have rubbed two pieces of very dry wood together until the heat lighted them, as many savages do at the present time; or they might have struck sparks from flint upon rotten wood and blown the spark into a flame. We may be sure that when once a fire was lighted they would take care it did not go out, and if they wanted to travel they would carry with them a piece of smouldering wood to light the fire again. I do not suppose that these pre-historic men were any more civilized than the savages of Australia and other countries, and I have often thought when looking at these savages that they live in almost exactly the same way as the earliest inhabitants of Europe did. They have the same shaped weapons and tools made of stone, and these are fixed to the handles in the same way. They have the same kinds of needles and fish-hooks made of bone, and they sew skins together with threads made from the sinews of animals. Thus we see men living now in many parts of the world who are quite as uncivilized as the old inhabitants of Europe, who lived perhaps thousands of years before the Egyptians and Assyrians.

These very ancient men knew nothing about metals. All their tools were made of flint, or bone, or stone, and they were of the rough shape you see in the pictures on the next page, and it is for this reason that this has been called the *Stone Age*. These were chipped out with great trouble and labour, and most of them were not even polished. With these they had to kill animals for food, to cut down trees, and fight against their enemies. The skeleton of a mastodon was found in the state of Missouri in America about thirty-five years ago with numbers of these flint arrow-heads underneath and near it. Perhaps it had been shot at with arrows, and when it died the flint points fell out of its decaying flesh. But it is not likely that these pre-historic men could have killed many such large animals, unless they caught them in pits covered over with branches of trees and earth,

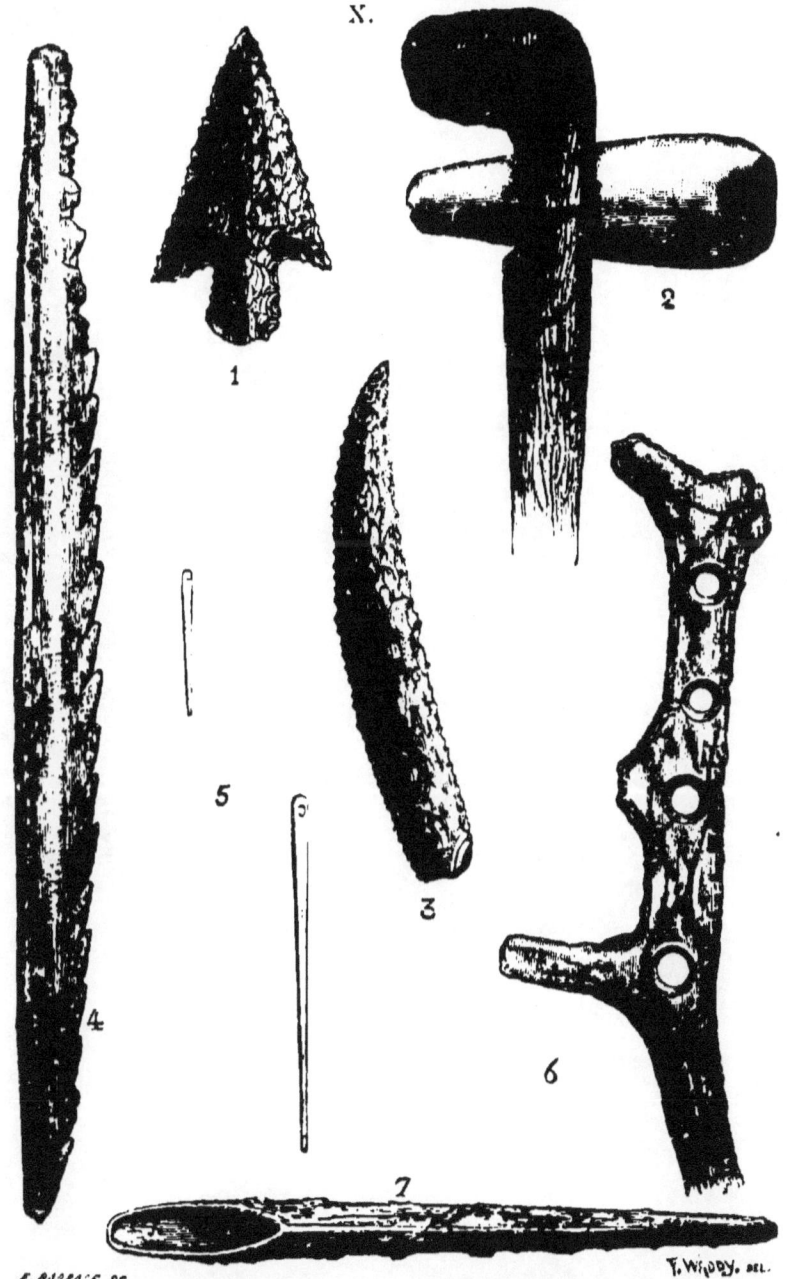

1. *Flint Arrow-head.*
2. *Stone Axe in handle.*
3. *Flint Knife.*
4. *Bone Harpoon.*
5. *Bone Needles.*
6. *Sceptre made of Horn.*
7. *Marrow Spoon.*

into which they might fall, as elephants are sometimes caught in Africa.

Nothing shows us so well the immense time which must have passed since the men of the stone age lived as that these flin weapons and tools are found nearly all over the world, in Northern Europe, including our own country, in Spain, France, Italy, Greece, Palestine, Africa, Japan, America, &c.; and yet none of the present inhabitants of these countries have any history or tradition of the time when they were used. Metals are now used instead, and there is no record of the time when flint only was known. We are quite certain however that the stone age men lived at the same period as the great animals of the Tertiary age, the mammoth, the mastodon, the woolly rhinoceros, the Irish stag, the cave bear, and others you have read of in former chapters, because flint and stone weapons are found in the same beds of earth with these animals.[1]

Suppose one of the present Indian or

---

[1] British Antiquities Room, upper floor, Middle and Upper Shelf-cases, Nos. 1, 2, and 5–12, flint and stone implements. Table-case B, horn implements from French caves and Swiss lake-dwellings.

African elephants with his rider were to fall into a river and they were to sink to the bottom and be covered with mud, and suppose his rider had in his pocket some of our sovereigns. If that elephant should be accidentally dug up thousands of years to come, when most likely all elephants will have died off the earth, people would know for certain, from the date and figure of the Queen on the money, that elephants were used by the English in this reign, even if all our books and monuments had perished, and a new people inhabited the Earth. Something of the same kind has happened to prove to us that the stone-age men saw the mammoth alive. In one of their graves there is a slice of a mammoth's great back tooth with a beautiful picture of the animal, with his bristly hair, scratched on the ivory, and there are also many of the flint and stone weapons which show that the skeleton in the grave was that of a primeval man. This little picture tells its tale more faithfully than any history. It is all the more certain to tell it truly because it was never *meant* to tell one. When that man was buried with this sign that he was a mighty hunter of the mammoth, or an artist, no one could imagine that he would ever be dug up to show us, who come so long afterwards, that he saw the mammoth roaming through the forests of the far away past. There can be no doubt that it is a very good drawing of the mammoth with its long turned-up tusks, like those in the picture at the beginning of the book.

In another place a picture of a fight between some reindeer scratched upon a piece of slate has been found. This was in a cave in France, and it, as well as the numbers of bones of these animals in the caves, shows that the reindeer, which now only inhabits the Arctic regions, must have been common then in France. You will see drawings of both these on page 135.

These primeval people built no houses. They lived in natural caves, and scattered the remains of their food about the floor, so that we know what they ate. Among the animals they used for food were the horse, the reindeer, the ox, the cave-lion and bear, the wolf, the hyena, the goat, the hare and several others, besides salmon and other fish. They were very fond of the marrow of the bones, which they cracked with stone hammers, and had little spoons made of bone with which to pick it out.

They had places for making flint weapons too. At Cissbury Camp, near Worthing, there is one of their old workshops. There are galleries dug into the chalk where they got the flints, and there are thousands of chips of flint lying about, with half finished arrowheads, and some of the tools they dug with. They had no spades or pickaxes; but they used the broad, flat, shoulder-blade bone of the ox as a spade, and the sharp brow antler of a deer's horn for a pickaxe, to get these flints out with. It must have been very hard work for them, because bone spades and horn pickaxes would soon wear out, and would not be nearly so useful as ours made of iron.

It is difficult to be certain how these stone-age people cooked their food. Of course they could have roasted it, and the half-burnt bones in some caves show that they did so; but in some caves in France there is not a single burnt bone to be found. In these French cave dwellings, too, there are

XI.

Picture of Mammoth Scratched on Ivory.

Fight between Reindeer Scratched on Slate.

no pieces of earthenware, as there are in some others; so that the people could not have boiled it, unless they had wooden pots and dropped red-hot stones into the water in them until the meat got boiled, as some savages do now. Or they might have cooked it under the hot ashes.

The people who used earthenware must have made more progress. It is easy to understand how they made this useful discovery. Suppose they had lighted a fire upon a damp clay soil, the earth would get baked hard and crack off in pieces, and they would see that this soil could be worked in the hands while soft into the shape of pans and dishes, which could be dried quite hard in the sun or baked in hot ashes, just as boys make clay marbles now. They could live much more comfortably even with these rough earthenware things, and cook their food more conveniently; but they still used the stone and flint tools and weapons, and iron was still unknown to them.

The people of whom I have been speaking are principally the men of the First Stone Age, when the art of polishing tools and weapons had not been found out. They simply chipped these things out of the flints and left them very rough; but the men of the next, or Second Stone Age, made great improvements. They ground their flint knives and axes with other stones, and rubbed them down to sharp edges and points, so that they must have been much more useful for killing and cutting up the animals they hunted. All their bone and horn tools are much better made, and sometimes ornamented prettily with marks cut upon them. The Second Stone Age men evidently wore clothing, most probably made of the skins of animals —for the long strips of bone with a hole at one end which you see in the picture could not have been used for any other purpose, except to draw threads through something. The threads were very likely either the sinews of animals pulled out of the flesh, or thin strips of their skins, or perhaps the inner bark of a tree twisted into a kind of string. In the colder parts of Europe and America these ancient people would need some protection from the weather. How then did the people of the First Stone

Age manage, if they had no bone needles, as I think they had not, with which to make clothing? They must have wrapped themselves in the skins just as they came from the backs of the animals.

It is not easy to be always sure, when we find a cave and all these relics of pre-historic man, whether the inhabitants belonged to the First or the Second Stone Age. Sometimes there are signs of polishing and grinding on the tools, and then we may suppose that men were gradually getting more skilful, until they finished off all their weapons beautifully. But there is such a very great difference in the perfection of these useful articles found in some places and those found in others that we have no doubt men made slow progress, from the rough or First Stone Age, to the polished or Second Stone Age.

In neither the first nor second stone period had men yet learned to build any kind of habitations. They lived in caves simply, like wild animals. On the banks of the river Vezère in France, which has cut its way deeply through the rock, there are some celebrated caves once inhabited by pre-historic men, and some of them are very large. They were most likely hollowed out in the cliff by water, and many generations of men lived here. In one of them four human skeletons were found, with plenty of stone and flint tools, besides the bones of the mammoth and lion, reindeer and other animals. The mammoth then as well as the reindeer lived at that time in the valley of the Vezère. There is no doubt that these caves were inhabited at separate times by people who used only the roughest and simplest stone tools, and by others who had made some progress and could polish their tools and make them of bone and could scratch pictures of animals upon slips of bone and slate. It is curious that all these drawings are side-view drawings, and they are only outlines, just like the drawings of children now, and the Esquimaux of the Arctic regions; because these people, although they were grown up, had not discovered the art of drawing in perspective and shading the figures. Still the pictures are wonderfully true to nature, and must have been copied from living animals. There is no earthenware in any of these caves, so that the useful art of making pottery had not been discovered, neither is there any in the caves in Switzerland, where the bones of the mammoth, lion, and rhinoceros are also found, and the tools and weapons are much the same as those in the French caverns. It is impossible to say whether the cave-dwellers of France and Switzerland lived at the same time exactly, but they were in about the same condition of civilization, and they must both have been quite familiar with the appearance of the mammoth and lion, and other animals, which are not mentioned in any history, however old it may be, as inhabitants of these countries.

A discovery has lately been made in France of a large cavern near Belfort, in the limestone rock, which has been covered up for ages. The quarrymen while cutting out the stone came upon a small opening leading into a very large cave, in which there was a great quantity of human skeletons and bones and some beautifully ornamented vases, polished stone bracelets, and a mat of plaited rushes. To these people, then, the arts of pottery and weaving were known, and this was probably one of their burying-places. They were evidently much more civilized than the ancient people of the valley of the Vezère; but this cave must also be of a great age, and its inhabitants have left no record of their history in any kind of writing.

Quite lately, too, we have learned something of the early races of man in Colorado. Many of the caves in that country have been altered and made more like regular houses, and some appear even to have been cut out of the rock entirely by human hands; and in the plains there are ruins of large cities.

Though still in the stone age, for all the weapons yet found among these ruins are of stone, the Colorado people were more civilized than the stone-age people of the Vezère caverns, because they had begun to build and knew how to make pottery. It is strange, too, that the present natives of Colorado are not so civilized as the early people, and if they have descended from them they have not improved, but rather the contrary. There are other caverns in various parts of the world containing these curious relics of races long since passed away, but some of the principal have been mentioned, enough perhaps to interest you and show you that men were living in Europe together with the large animals of the Tertiary period, and that they had made very little progress in the arts and manufactures, and had not even begun to build the roughest houses.

In many parts of the world even now there are savages nearly as uncivilized as the cave-dwellers of Europe were then. When Captain Cook visited New Zealand, more than a hundred years ago, the natives there had nothing but stone and bone tools, very like those found in the European caverns, and the inhabitants of some of the islands in the Pacific Ocean still use stone axes and hammers and bone needles.[1] Captain Moresby, too, who made a voyage to the south-east coast of New Guinea a few years ago, tells us that the natives have beautiful stone axes, but they were so ignorant of the use of iron that they refused to give him one of their

[1] Examples of stone implements of New Zealanders in Ethnographical Room, Cases No. 45-48, upper floor.

stone axes for a new iron hatchet which he offered them. No doubt the stone weapon cost a great deal of labour and patience to make, and perhaps the iron one was made by machinery in a few minutes, and was really more useful, but the native had proved his own axe and knew nothing of the iron one, so that it is no wonder that he refused it. But what a history these two axes tell—the stone and the iron! The stone shows us man in his childhood, and the iron man in his manhood, and what an immensely long time there is between the two. How much thought, and trial and failure, and patience and industry, were spent by mankind before the stone axe grew into the iron!

In Europe man has long since grown out of his childhood, but in many parts of the world he is no more civilized than the men who saw the mammoth crashing through the forests of England and France, and heard the lion roar at night on the banks of the Thames, and watched the hippopotamus swimming across the river at Westminster. It is most likely, then, that Europe and parts of Asia and America were inhabited long before those places where men are even now in the stone age—such as the islands in the Pacific Ocean, New Guinea, Australia, &c.

What a life the pre-historic men of Europe must have lived! Here they were surrounded by huge dangerous animals, and had no means of protecting themselves against them but with these rough stone weapons. Where London now stands with its miles of streets and busy life there was a mighty forest, and the mammoth and rhinoceros tramped through it by day, and the lion and hyena hunted the deer at night. When the pre-historic men came down to the banks of the Thames in the day-time to spear salmon, they saw the hippopotamus plunging about in the water among the rushes, sweeping the long grass into their wide mouths, and swimming from side to side with their young ones perched upon their necks. It must have been a grand sight, but a fearful one too, and it is no wonder that men thought the caves the only safe places to live in.

Sometimes in India the elephants come into the villages at night and throw down wooden houses and kill people, and they are very much feared, so that we can suppose how much more terrible the mammoth might have been to the uncivilized cave-dwellers. If they shot at him with the flint-pointed arrows they could scarcely hurt him, and it is more likely that they got out of his way as quickly as possible whenever they met him, and took good care never to interfere with the lion and rhinoceros.

THE LAKE-DWELLERS.

Among the earliest inhabitants of Europe, there were some who did not live in caves; but I think they must have lived a long time after the cave-dwellers, when they built their houses out in the middle of the lakes. These houses were built in a very curious way, and the remains of them have been discovered in Ireland and Scotland, Switzerland and other countries. The people carried quantities of stones, and earth, and sticks out into the lake and let them sink to the bottom. Then when they had piled up enough to make an island, they laid wood across and set up theirhuts, and lived there surrounded by water. These were very poor houses of course; but

XII.

Lake-Dwellings.

when men had begun to build for themselves, they would find how much more comfortable they were than in damp and dark caves. They must have had some kind of boats or canoes, or they could not have passed between their lake-dwellings and the land unless they swam to them ; but I do not think that any of these boats have been found. Perhaps they were made of the dried skins of animals stretched over wooden frames, as I have seen savages make boats.

There was another way of building these lake-dwellings, and a better way too. Long poles were driven into the earth at the bottom of the water, and when the builders had got enough of these together they laid other poles across them, and built their huts on this floor above the water. People are living now in much the same way near the Orinoco river in South America, in New Guinea, and in Central Africa.[1] The land all round is covered with water from the overflowing of the rivers, which are very large, and the huts are built up on these poles out of the way of it. The lake-dwellers of Europe would thus be safer in their

[1] In Lake Mohrya. *Across Africa*, by V. L. Cameron.

houses from dangerous animals than if they were on land. They were more civilized than the cave-dwellers, but still a great many of their tools and weapons were of stone and bone ; yet we know that they had made wonderful progress, because they had learned to make pottery, and even to weave cloths out of hemp or flax. They had most likely begun to plant and cultivate the land, too, for corn is found about these dwellings, and the bones of domestic animals are very numerous. They had left the cave-dwellers a long way behind in many things, in wearing artificial clothing, in cultivating the land, and in keeping domestic animals; but their implements—that is, their weapons and tools—were not much improved, and were very much like those of the cave-dwellers, though better finished and more polished than some of theirs.

But not all the articles used by the lake people were of stone and bone. Some of those who lived in the Swiss lakes had ornaments, such as bracelets and hair-pins, made of the metal called bronze, and no doubt they made spear-heads of the metal, because they would look to usefulness before ornament.

THE LAKE-DWELLERS. 151

Now you see how these people seem to have lived : first the old stone age men, then those of the newer or polished stone age, and lastly the lake-dwellers. The people of both the first and second stone ages certainly saw the mammoth, hippopotamus, rhinoceros, lion, and reindeer alive in France, Switzerland, and England; but when the lake-dwellings were built, all these animals, except perhaps the reindeer, had died, and most of the animals were the same as they are now. None of these people have left us any kind of history whatever, except that which their simple works tell us, their flint and bone weapons, and their dwellings. They have set up no gigantic monuments like the Egyptians or the Druids. They thought of no men to come after them who would take an interest in their ways; but it is fortunate that what they did make was of such lasting materials as stone and flint, or we should have known next to nothing about their lives.

It is impossible to say how many thousands of years may have passed before the rough stone weapons were replaced by the polished stone, or the cave was exchanged for an artificial house in a lake; but you must feel in your minds that the time was immense, and the more we study the ways and works of pre-historic man, the more certain we become that it is longer than the whole time that has passed since men first began to use any kind of writing.

KITCHEN-MIDDENS.

I dare say you have seen untidy people in country places, and even in towns, throw oyster-shells and broken dishes and dirt outside their doors until quite a heap is formed. This is called a "midden," and the habit of doing this is a very old one. We learn just a little more of the history of man from great middens made by ancient people in several countries. They were first discovered in Denmark, and since then they have been found in Scotland, Brazil, and New Zealand. They are sometimes very large, and must have been used by the whole village as places to throw the refuse of their cookery in. When these heaps have been dug into all sorts of things have been found in them—the shells of oysters and mussels, bones of fishes, birds, and animals, pieces of broken earthenware, little ornaments, stone axes, arrow-heads, wood ashes, burnt bones, and other odds and ends. In Brazil many of these kitchen-middens are on the sea-shore, and it seems as if the people who made them came there to live on the shell-fish, for the shells are the same as those living in the sea close by now. In New Zealand the middens contain many of the bones of the Moa, which was described in "The Animal Part," and has now perished, and these are cracked in such a manner that the people evidently wanted to get at the marrow in them, and it shows too that this gigantic bird was common in New Zealand then. The midden makers seemed to have lived in the open air, and wherever food was most plentiful. Perhaps they built huts of the bark and small branches of trees like the Australian savages, but such houses would not last. We only know of the life of the midden makers from these heaps. Their weapons are of the same kind and pattern as those of the Second Stone Age, but they had learned to make rough earthenware dishes and basins, and some pieces of a woven material have been found, and pieces of wood and bone worked with a little skill. Whether they lived after or before the lake-dwellers I cannot say, but I should think about the same time.

These pre-historic people, nevertheless, were not always thinking of making things which were useful. They thought too of making ornaments, many of which are found in their dwellings and graves. Like ourselves, they had an idea that little trinkets improved their appearance. In one grave a skeleton was found with a small pile of shells under its neck, which no doubt had been strung together as a necklace, and when the string rotted the shells parted and fell in a heap under the head, to be a memorial of that ancient man or woman's possession of the same feelings as our own. Various little articles, too, found about the lake-dwellings show that people liked to decorate themselves.

We shall never know what language they spoke, but they must have been able to tell their thoughts to one another. It was most likely a simple language with few words as names for things and a simple grammar, like the language of savages, because they had not so many things to talk about as we have. The names of animals would perhaps be imitated from their cries and the noises they made. These cries would be among the most familiar sounds to them, and when they wished to speak of some animal the simplest way would be to imitate the noise it generally makes. If we think of our own language, we shall see how very likely this was. We have many such words. We teach our children the names of animals by the sounds they make. The dog we call "bow-wow," the cow "moo-moo," the duck "quack-quack," and many other names of the same kind which you will think of yourselves. At the present time even the name by which the Egyptians call the donkey has almost exactly the same sound as our "hee-haw." This trick of doubling or repeating the sound, too, is very common among savages, who are as far behind us as the pre-historic men were. The natives of Australia give these double names to a great many animals and things, and sometimes do the same with English words. They call fish "ningy-ningy," and a certain tree the "bunya-bunya," and their language is full of such words. But it is not only the names of things which have been made in this way. Verbs as well as nouns have grown up thus. When we whisper to one another, that word imitates the low sound we make.

I shall leave you to trace the natural origin of the following words, and think how much of man's spoken language is taken from common sounds. Thus we have roar, shriek, whistle, hiss, sigh, sing, ring, thump, bump, clash, clang, bang, twang, clap, smack, slap, smash, swish, swirl, gong, thong, boom, bellow, batter, chatter, clatter, snap, snip, whip, gurgle, shiver, quiver, rumble, roll, rattle, prattle, and a hundred more. Words thus derived from familiar sounds abound in all languages, and they, no doubt, are the easy steps by which men climbed to a more complicated speech. The earliest men must have been obliged to pay great attention to animals and birds, which have voices of their own; for to hunt and catch them was the principal occupation of their lives; therefore, when speaking of them to one another, they would naturally call them by names resembling the sounds they made. Our verbs "to squeak" and "to squeal" are certainly taken from the cries of animals when in pain; but I have said enough to show you how language grew up among pre-historic people.

We do not know for certain that they had any musical instruments, but they would hear the sighing of the wind among the trees, and it would almost certainly be found out that blowing down a hollow stick or reed, open at one end and closed at the other, would make a whistle; but if they used any of these things they would not last like the stone tools, and have decayed away; and we do know that they had begun to draw upon such imperishable materials as bone and slate.

There is a very interesting specimen of a human fossil in the British Museum, which you ought to go and see, if you can; but in case you are not able there is a drawing of it on page 159.[1] This specimen was brought to England about the year 1814. Others like it have since been found imbedded in the hard breccia limestone rock at the same place on the shore of the island of Guadaloupe.

[1] At the end of Room VI., opposite the door, North Gallery.

The skeleton most likely was that of a woman, from the shape of some of the bones, and most probably was of the race of Caribs, of whom there are none living now. Perhaps this was originally a burying place of the ancient inhabitants of the island, and when the sea washed the small broken pieces of shells and corals over it (all of which contain lime) they hardened into breccia rock, and the skeleton became completely imbedded in it. This must have taken a very long time, at all events; but I do not think the Guadaloupe fossils are as old as the people who lived in the caves in France. Some little ornaments and articles of human workmanship are found with these skeletons, which show that the people to whom they belonged were still in the Stone Age. There is very little to judge from when we wish to get some idea of the time these fossils have been in this breccia: but at this particular place the rock is formed pretty quickly, as we can see; and it is quite likely that these skeletons were buried there long after the mammoth, rhinoceros, and hippopotamus died out of Europe. However, they are the most com-

XIII.

The Guadaloupe Human Fossil.

plete specimens we have of any fossil human beings. In looking at the drawing you will see the leg bones and hips, part of the back-bone, the ribs of one side, and an arm bone; but you see no skull, because the bones of the skull are very thin, and have become crushed down into the limestone. In one of these fossils, which they have in Paris, taken from near the same place, the bones are much more distinct, and part of the lower jaw with some teeth in it can be seen. These fossil men no doubt lived before the period of written human history began; but they are not considered to be at all the oldest of pre-historic men.

Two periods in the life of mankind followed all these long-lost and forgotten people, and they are called the Bronze Age and the Iron Age; but now *history* comes in, and there are plenty of old records and books to tell you about these. Bronze is a mixed metal of copper and tin, and it was used by the oldest nations who have left any histories —the Egyptians, Hebrews, Greeks, and Romans. It was better than stone because it could be made sharper and would not chip, and swords and armour, vases, axes, hammers, needles, &c., were made of it.[1]

The Stone Age is beyond all history, the Bronze begins with it, and the Iron Age began at some distant time before the dawn of authentic history. Thus we are told, in Genesis iv. 22, that Tubal Cain taught people to make it. It was used also by the Egyptians for perhaps 2,000 years before the Christian era; but the real Iron Age is that in which we are living now. We can, indeed, make all metals much better than any of the older nations.

But there is a wide gap between the time when people left off using stone and discovered bronze and iron; and if one of the Druids could come to life he might help us to fill it up, because those old British priests had many secrets, which they told to one another from generation to generation.

If the Spanish conquerors had not destroyed the civilization of Mexico and Peru, we might know something of the discovery of the metals there, and the people of India and China must have used them long ago; but

---

[1] See examples in the Bronze Room, upper floor, British Museum.

the first use of metal in any country where it was found out would most likely be before the people had begun to put their language into any kind of writing, so that the time would be forgotten among the many scraps of lost knowledge which we have tried to collect from the remains of the industry of pre-historic man.

We have seen how much these ancient people differed from us in their civilization, and how far they were behind us in everything; but we must not suppose that they were very different in bodily size and shape. Some of their skulls might have belonged to a philosopher, or they might have contained the thoughtless brains of a savage. The skulls from the Cromagnon and Engis caves are quite equal in size and shape to those of several uncivilized, and even of some civilized races of the present time, and there are people in all large cities whose heads are not better formed. Though the outward signs of their civilization then were so different from ours, it is not certain that their mental capacity was much less.

A race possessing considerable civilization may, we know, pass away, as the

Assyrians and the Pyramid builders have. In one of the Pacific islands—Easter Island—a thousand miles from the nearest land, there are hundreds of carved images of stone, fifty or sixty feet high, and weighing perhaps a hundred tons each. The people who made these must have been very numerous and must have had considerable skill. Yet they have passed away. The arts of Nineveh and Babylon have only lately become known, so that, you see, the works of a race may easily become hidden from us who follow. Quite lately, too, the works of a partly civilized people have been discovered in Ohio in America. There are there hundreds of mounds and earth embankments forming fortified camps. Some of them are several miles round, and they could only have been made by a very numerous and intelligent people who knew something about geometry; for the circles, squares, and angles of these earthworks are quite as correct as we could make them. Among the multitude of things found here are copper tools made by hammering, ornamental pottery, silver beads, plates of mica with scrolls and designs engraved on them, and carefully carved pieces of stone. These carvings are most curious and excellently finished. They represent human heads and many animals, such as the bear, otter, wolf, beaver, raccoon, frog, rattlesnake, heron, crow, &c. A people, then, who could do these things and took pleasure in doing them must have possessed great intelligence and a knowledge of things far beyond a simple state. They even had religious ideas, such as they were, for they had places for sacrifice. All their works are now overgrown by forests, but it is impossible to mistake them; yet the native Indians of Ohio living now have no idea that such a people lived in their country before them, and no tradition at all about a people whose civilization was so far superior to their own.

We may come nearer to our own times, and look at the Assyrians and Egyptians. Until quite recently nothing was known about the Assyrians except what could be learned from the few references made to them in Scripture and some ancient writers; but Mr. Layard dug up their cities, and found that they possessed the arts of building, sculpture, working in metals, and a written language. All this was buried under the sand of a desert! Then there is the great Pyramid of Egypt, built in a way that we could not surpass, and with much knowledge of geometry and other sciences.[1] The men who designed and constructed these works could not have lived among a half-barbarous people; and as these are the highest works of the people, how much there must have been that went before, of which there is no trace now, when Assyria and Egypt were in *their* age of stone axes and flint arrow-heads.

I do not think that the Stone-Age men of Europe were nearly so civilized. At all events, they have not left any such imperishable monuments as the gigantic images of Easter Island, the earthworks of the Ohio people, or the sculptures, writings, and buildings of the Assyrians and Egyptians; but they might have been more civilized than they seem to have been from their simple weapons and tools. They might have made many

---

[1] Built of nummulitic limestone, composed of shells of foraminifera. See Case 15, Room V., North Gallery.

things which were perishable, and have been destroyed by time—things which would have given us a higher belief in their intelligence and civilization.

The past history of the human race may be compared to the rise and fall of the tide. Wave after wave has risen higher and higher on the everlasting shore of Time, and when the tide was at its highest it has fallen again slowly, to rise again and again in the same way through many ages. We know that man may rise slowly from a simple condition to much civilization and power, and may again sink back almost to barbarism, as has been the case with the people of whom we have been speaking, and then again a new civilization may grow up. It is possible that all now savage nations are the sinking descendants of some, in comparison, once civilized people. Modern nations are taking up the ground of savages all over the world, and soon there will be no trace of these simple people. Thus it may have been with mankind throughout all the time during which they have occupied the earth, and so it may be perhaps again.

CONCLUSION.

I HAVE now put "The Puzzle of Life" together as well as I can, and there is not much more to say. You must do the rest for yourselves by going to the Museums, where all the pieces are collected, and seeing them with your own eyes. When you stand before these silent witnesses to the great age of our Earth, and all that is on it, you will feel how wonderful the story they tell is. They have no words to speak to you, but there is a power in your own minds which interprets their history through your own thoughts. They are only lumps of rock and lifeless bones, but they seem to say to you, "We are living again now, because we are teaching you a lesson which the great Builder of this Universe wishes you to learn from us. There is not a stone or fossil among us but it has its tale to tell—a tale of time and tide, and long past ages, and innumerable changes, and a life that was, and progress from a lower to a higher existence. We have obeyed the same eternal laws of one Creator from the beginning, as all things will to the end of time. We have opened the great Book of Nature from the first page of the 'life-dawn animal' to the last, on which the hand of the Almighty has written the name of Man—his most perfect work. We, you, and all things which have lived and will live, have bodies made of particles which will be returned to the Earth, no single atom of which has been destroyed since the first, but has been fashioned over and over again into innumerable forms of tree and flower, of gossamer-winged insect and towering mammoth, throughout the long ages in which our Globe has known day and night, cold and heat, summer and winter."

There is nothing sad, if we look at it rightly, in this constant succession of life and death. It is

> A moulding
> Of forms, and a wondrous birth,
> And a growing and fair unfolding
> Of life from life, and life from death.

> For death, a mother benign,
> Transformeth but destroyeth not,
> And the new thing fair of the old is wrought.
> G. F. ARMSTRONG.

Is it not worth while then to listen to these stories of the Earth—to spell them out for ourselves? They are written everywhere,—in the mountains and valleys, the rivers and seas, on the hard faces of granite cliffs, on the rounded pebbles of the sea beach, and even in the finest dust of the roads. We have not to go far to hear them: every foot-step on the ground covers a chapter great or small in the universal history, and the stone walls of our houses could speak with ten thousand tongues of all they witnessed in their long life on the floor of an ancient ocean.

We can scarcely have a more pleasant occupation and greater interest than in searching for and putting together the pieces of this wonderful and beautiful puzzle, and in doing our utmost to "Summon from the shadowy Past the forms that once have been."

INDEX.

A

AGE of bronze, 161; of iron, 161; of reptiles, 81
ALEPH, 125
AMBER, 69
AMMONITES, 90, 97
ANIMAL PART, the, 77; animals of coal period, 71
ANTS, white, 61
ARCTIC climate, 67; expedition, 67
ARCHÆOPTERYX, 91, 93
AUSTRALIAN savages, 127

B

BABYLON and Nineveh, 164, 165
BEAR, grisly, 106
BEGINNING of life, 58
BIRD forms, earliest, 89; reptiles, 85
BLACKLEAD, 58
BOULDERS carried by ice, 48
BOGWOOD, 70
BOILING springs, 54
BRONZE, age of, 161, 162; implements in British Museum, 162
BRIGHTON Downs, 99
BURNING mountains, 19

C

CALAMITES, 42, 68
CAÑONS of Colorado, 8
CAVES of Engis and Cromagnon, 163; near Belfort and of Switzerland, 141; of the Vezère, 139
CETIOSAURUS, 86
CHALK, nature of, 26; pits, 20; ammonites and foraminifera in, 27; period, 95; under the ocean, 29, 99
"CHALLENGER" expedition, 27

CHANGES have been gradual, 43
CISSBURY camp, 134
CLAY, London, 21, 22; and mud, 33
CLIMATE, Arctic, and of coal formations, 67
CLUB-MOSSES, 61
CLOTHING, 138
COAL beds, 31; in Arctic regions, 67; plants of the, 63; is fossil wood, 73; is sunlight compressed, 30
COLORADO, the people in, 142
COMPRESSED PLANTS, 15
CONCLUSION, 168
COOKERY, 137
CORALS, 78
CREATION, the plan of, 117
CRETACEOUS period, 96
CROMAGNON and Engis, caves of, 163

D

DAWN of life, 56; plant, 59
DENUDATION, 49, 50
DINORNIS, specimens of, in British Museum, 116
DINOSAURUS, 89
DINOTHERIUM, 114
DRAWINGS, pre-historic, 135
DWELLINGS and food of men, 137

E

EARLY histories, 123; plant life, 59
EARTH, early history of, 1, 2, 3; interior of, 18; intense heat of, 24; climate of, 48; not yet fit for man, 75; 'foraminifera earth,' 30

INDEX.

EARTHQUAKES, 18, 19
EARTHWORKS of Ohio, 165
EASTER island monuments, 164
EGYPT, monuments of, 266
EODENDRON, 59
EOPHYTON, 59
EOZÖON, 57, 77

FIRST weapons, 121
FISH-LIZARDS, 85
FISHES, fossil, 71
FLINT, origin of, 14; in chalk, 96; weapons, where found, 131; tool manufactory, 134
FORAMINIFERA, 20; 'foraminifera earth,' 30; drawings of, 97; specimens of, in British Museum, 99
FORESTS under the sea, 75, 76
FOSSIL, derivation of, 10; plants, 61; sunlight, 73; footprints, 83; human, 157, 159
FOOD and dwellings, 137
FOOTPRINTS, fossil, 83
FLYING reptiles, 89

GEOLOGICAL part, 17
GEOLOGY, derivation of, 19
GEYSERS, 54
GIGANTIC animals, 101; birds, 115
GLACIERS and icebergs, 47
GRANITE, raised, 23; appearance of, 24
GRAVEL, &c., 35
GREAT IRISH STAG, drawing, &c., of, 107
GUADALOUPE human fossil, 157

HEAT of the Earth, 3, 18
HEBREW letters, 125
HESPERORNIS, 92
HIPPOPOTAMUS in England, 105
HISTORIES, early, 123
HUMAN part, the, 120; fossils, 157

ICE age, 45; more than one, 48
ICEBERGS and glaciers, 47
ICHTHYORNIS, 92
ICHTHYOSAURUS, 85
IMPLEMENTS, flint and stone, in British Museum, 131; bronze, 162
INDIA, elephants in, 145

INSECTS in coal forests, 64
IRISH stag, 107
ISLANDS appear and disappear, 39

JET, 69
JURASSIC age, 89

KANGAROO, fossil, 115
KITCHEN middens, 152

LABYRINTHODON, 2
LAKE dwellers, 146; dwellings in Europe, Africa, Asia, and New Guinea, 149
LANGUAGE, origin of; and of prehistoric man, 155
LAURENTIAN rocks, 57
LENA river, mammoth found, 102
LIFE, the dawn of, 56; 'life-dawn animal,' 57
LIGNITE, 69
LION, English sabre-toothed, 106

MAMMALIA, 102
MAMMOTH, 49, 102-3; bones of, in Siberia, Asia, North America, &c.; drawing of, on ivory, 135; in Essex, 104; skull of, in British Museum, 104
MAN and his works, 121; his earliest inventions, 122; mammoth, mastodon, reindeer, &c., contemporary with, 116; pre-historic, 127, 131; dwellings and food of, 137
MARSUPIAL animal, 95
MASTODON, 102; in Europe, America, India, &c., 104; in Missouri, 128; skeleton of, in British Museum, 104
MEGALOSAURUS, 89
MEGATHERIUM, in South America, 110; drawing of, 112; account of, 113; skeleton of, in British Museum, 113
MEXICAN writings, 124
MIDDENS, kitchen, 152-4; makers, life of, 153
MOA, 115-16
MONKEYS, fossil, 102; at Gibraltar, 102

INDEX. 173

MONUMENTS of Easter Island, 164;
   of Egypt and Assyria, 166
MOUNTAINS, burning, and covered
   with snow, 19
MORESBY, Captain, in New Guinea,
   143

NEW GUINEA, stone age of, 143
NEW ZEALAND dinornis, 115; moa,
   116; stone age of, 143
NINEVEH and Babylon, ruins, &c., of,
   164, 165
NORWAY, raised terraces of, 38

OHIO, earthworks of, 165
OOLITE, 41, 86
ORIGIN of language, 155

PAPYRUS writings, 125
PARIS, built of shells, 100
PARTS, the, are called fossils, 11
PAST life, the signs of, 13
PEAT, 70
PLAN of creation, 117
PLANTS of coal forests, 63
PLESIOSAURUS, 85
POTTERY, 141, 142
PRE-HISTORIC art, 133; drawings,
   135; man, 127, 131; weapons and
   tools, 129
PTERODACTYL, derivation of, 89;
   description of, 90
PUZZLE, the framework of, 1-16;
   parts of, where found, 5
PYRENEES, when raised, 100

RAIN-drops, marks of, 84
REINDEER, drawing of, on slate, 135
REPTILES, the age of, 81
RHINOCEROS in England, 105

ROCKS, raising of the; how placed,
   21, 25; carried by ice, 48

SANDSTONE, formation of, 25, 26;
   Old Red, 62, 81; New Red, 77
SLATE hardened mud, 15
SPONGES, 15, 78
STAR-fish, 78
STONE age, 128; first stone age, 137;
   second, 138; of New Guinea and
   New Zealand, 143, 145
SUBSIDENCE, 37
SUCCESSION of formations, 41, 42
SUCKLERS, 102
SUNLIGHT, fossil, 73

TERTIARY period, 34, 100
TIME, the work of, 167
TOOLS, polished and rough, 139
TRILOBITE, 78

UPHEAVAL and depression, 36, 38

VEGETABLE part, the, 56
VERTEBRATA, 101
VOLCANOES and earthquakes, 19

WATER, a powerful tool of Nature,
   34, 45; thrown out of the earth,
   54
WEAPONS, early, 121; and tools,
   where found, 131
WHALES, 101
WORLD, early history of the, 3, 4;
   size and shape, 17; materials of,
   17; heat of, 18
WORK, the, of time, 167
WRITING, origin of, 123; Mexican
   Egyptian, and Assyrian, 124, 125;
   on papyrus, 125; by signs, 125

LONDON: PRINTED BY
SPOTTISWOODE AND CO., NEW-STREET SQUARE
AND PARLIAMENT STREET

OPINIONS OF THE PRESS.

'The present little work, which is specially addressed to children, is written in so pleasant and easy a style, and its descriptions of life on the earth are on the whole so simple and accurate, that we can heartily recommend it to the attention of those who seek such a guide. The illustrations are good, and the general appearance of the book such that it may compare most favourably with other primers of geology.' GEOLOGICAL MAGAZINE.

'Written in clear and simple style, especially attractive to children. It includes an account of pre-historic man, and shows in many other ways that the writer is familiar with some of the latest phases of geological thought.' ACADEMY.

'The avowed object of this charming little book is to place the results of these researches within the grasp of children, by presenting them in language at once clear, simple, and winning......In this hard task Mr. NICOLS has succeeded admirably, without resorting to that base subterfuge—the attempt to clothe instruction in the guise of fiction......This is true education, for it teaches children first to observe and then to reason......Though the style of this delightful book is simple and childlike, it is as far as possible removed from being childish.' PALL MALL GAZETTE.

'The language is plain, the descriptions are lucid, the illustrations apt, and the broad facts of the science are very correctly stated. The work, too, is free from all attempts at fine writing.........We wish the book success as at any rate an attempt to lay before the young fact instead of fiction.'
QUARTERLY JOURNAL of SCIENCE.

'The book is a successful attempt to explain the simplest facts of geology, and of the succession of life on the earth.'
WESTMINSTER REVIEW.

'The idea is a happy one, and will recommend itself to children; and we are bound to say that Mr. NICHOLS has carried out his idea remarkably well, and produced a work which will do much to spread sound notions upon the gradual development of our earth and its inhabitants to the condition in which we now see them.........We can safely recommend Mr. NICHOL's little book as one that will have a most beneficial effect in opening the minds of its young readers.' POPULAR SCIENCE REVIEW.

Opinions of the Press—*continued*.

'This is a good little book, cleverly written by an able geologist, and well adapted for children. We can recommend the volume as a present to any intelligent boy or girl.' LANCET.

'This book appears to be, in style, language, and scope, eminently adapted for its purpose, which is to awaken among the little folks an interest "in the history of life upon the earth," and "give them the taste for more extended study in after years."'
ILLUSTRATED LONDON NEWS.

' "Though these pages are designed for young persons," says the Author, "other readers, perhaps, who are not familiar with the subject, may find some interest in them, if they are not deterred by the necessarily simple style,"—which, we venture to say, they most assuredly will not be......To many grown persons, therefore, as well as their descendants, will this book be a great boon, which, if they are at all liberal-minded, they will advocate as well as appreciate. ......Like the Science Primers of Professors Huxley, Roscoe, Balfour Stewart, &c., if duly read and weighed, it will tend to unravel and sweep away a deal of baneful superstition.' LAND and WATER.

'That Mr. NICOLS has succeeded in the object he proposed to himself may be safely affirmed. He has done his work briefly and lucidly, and has produced a book capable of arresting the attention, not only of children, but of those from whom they receive their earlier lessons.' The COUNTRY.

'A perfect "Open Sesame" for young scientific students, and so cleverly composed as to make students of those who are not scientific: not merely the young, but older people too. Mr. NICOLS thoroughly understands his work.' NOTES and QUERIES.

'Easily and attractively written for young people......The treatment of so wide a subject, and the condensing it into a volume of 150 pages is no light task. We can, however, congratulate Mr. NICOLS upon having accomplished it in so judicious, perhaps, better still, so suggestive a manner; and we have no doubt that his little book will become a well-worn favourite in the hands of all thoughtful and intelligent children who may be so fortunate as to possess it.' ENGINEER.

'The manner in which the pieces of the puzzle—fossils—are found, put together, and interpreted, is related in language readily understood by children; the description of the vegetable, animal, and human parts being peculiarly interesting. The illustrations are the best of the kind with which we are acquainted......We strongly recommend it.' SCHOOLMASTER.

OPINIONS of the PRESS—*continued.*

'It is the puzzle as to the history of life on the earth unravelled in a manner to interest and enlighten the minds, and to develop the observing and reflecting faculties of children......The results of costly and laborious investigations in many different branches of science are concentrated in these free and easy lessons or colloquial lectures to young children......Calculated to arouse an interest in all but the dullest and most indifferent juvenile minds......Will be found invaluable to teachers and a great help in the rational cultivation of the intelligence of the rising generation.'
SCHOOL BOARD CHRONICLE.

'The statement of these facts, though made with all the sobriety due to a scientific discourse, has all the interest of a story for the young; and the narrative, if we mistake not, will interest other readers than those for whom it is primarily written. A word of commendation must be given to the illustrations, which are exceedingly well drawn.' EDUCATIONAL TIMES.

'To place the "simple truths of science" in rivalry with fairy tales and merry picture-books is not so hopeless as at first sight may seem; and certainly the simple, attractive style in which the marvels of the physical world are here set out must not only interest, but charm every bright child of eager intellect. Simplicity is observed to the utmost, but it is the simplicity of truth, so that the child is not interested at the expense of having afterwards to unlearn what he has read or listened to.'
LIVERPOOL WEEKLY ALBION.

'Mr. ARTHUR NICOLS has attempted a task which at first sight seems extremely difficult, but which he has successfully achieved. ......Children can scarcely help understanding and being interested in the wonderful story of the earth's crust, and of past organic life upon it, which he unfolds. There is nothing childish about his style, yet he writes with perfect simplicity......A better book to put into the hands of thoughtful children, or for use as a text-book by persons engaged in the private tuition of the young, it would be difficult to find.' The SCOTSMAN.

'Facts are stranger than any fancies which emanate from the writers of even fairy tales, and when they can be brought home to youthful students by ocular demonstrations the facts are invariably preferred to the fancies......The illustrations which adorn the book are well drawn, and sufficiently numerous for the purpose......The Author is a genial and reliable guide to a solution of the puzzle of life.' ENGLISH MECHANIC.

London, LONGMANS & CO.